Theory of Satellite Geodesy

Applications of Satellites to Geodesy

William M. Kaula

University of California at Los Angeles

DOVER PUBLICATIONS, INC.
Mineola, New York

Copyright

Bibliographical Note

This Dover edition, first published in 2000, is an unabridged, slightly corrected republication of the edition first published by Blaisdell Publishing Company, Waltham, Massachusetts, in 1966. A new Preface to the Dover Edition has been added.

Library of Congress Cataloging-in-Publication Data

Kaula, William M.
Theory of satellite geodesy : applications of satellites to geodesy / William M. Kaula.—Dover ed.
p. cm.
Originally published: Waltham, Mass. : Blaisdell Pub. Co., 1966.
Includes index.
ISBN 0-486-41465-5 (pbk.)
1. Satellite geodesy. I. Title.

QB343 .K34 2000
526'.1—dc21

00-064345

Manufactured in the United States of America
Dover Publications, Inc., 31 East 2nd Street, Mineola, N.Y. 11501

PREFACE TO THE DOVER EDITION

There have been significant changes in satellite geodesy since 1966, with the great improvements in tracking and computing capabilities. A history of the advances in methods and results is given in the article "Gravity field determination and characteristics: Retrospective and prospective" by R. S. Nerem, C. Jekeli, and W. M. Kaula, in the *Journal of Geophysical Research 100*: 15,053-15,074, 1995.

Despite the many changes in technique, *Theory of Satellite Geodesy* is, I am told, helpful in gaining insight to the geometry and dynamics, and thus useful as a textbook.

W. M. Kaula

PREFACE

THE FIRST PURPOSE of this text is to demonstrate the application and development of familiar physics—Newtonian gravitation—and familiar mathematics—Euclidean geometry—in a particular environment: the earth. The second is to collect and explain some of the mathematical techniques developed in recent years in order to utilize artificial satellites for geodesy.

To make this book as useful as possible for the first purpose, it is assumed that the reader has completed only the pertinent parts of a first-year course in physics and a first-year course in calculus. Thus, although it is assumed that the reader is familiar with the fundamentals of potential fields and analytic geometry, full explanations are given of certain mathematical techniques, such as spherical harmonics and matrices, that are necessary to apply these fundamentals in the context with which the book deals.

The second purpose may interfere with the first in that it introduces more complications than are necessary for an understanding of the physical principles involved. However, the fact that such complications occur in applying simple physics and the fact that certain mathematical techniques are valuable because they cope most effectively with complications are perhaps the most important lessons for the student.

In writing this book I am indebted to the following persons: for comments on earlier drafts, to Robert G. Wilson and Bernard F. Cohlan at the University of California in Los Angeles and to C. A. Whitten, Martin Hotine, and others at the United States Coast and Geodetic Survey in Washington, D.C.; to Robert H. Gersten of Aerospace Corporation, Los Angeles, who contributed greatly by carefully reading the final draft; and to Elizabeth Doty for her diligence and patience in preparing the typescript and its multifarious corrections.

W. M. KAULA

CONTENTS

TABLE OF SYMBOLS

Symbol	*Definition*	*Dimension*	*First Used in Equation*
A	Cross-sectional area	L^2	(3.152)
A	Azimuth		(4.18)
a	Semimajor axis of ellipse	L	(1.36)
a_e	Mean equatorial radius of earth	L	(3.53)
b	Semiminor axis of ellipse	L	(1.36)
$\mathbf{b}_T$	Vector referring to earth-fixed camera axis	L	(4.18)
$\mathbf{C}$	Matrix of coefficients of corrections to observations		(5.41)
C_D	Shape factor		(3.140)
C_i	Calculated value of quantity observed		(4.32)
C_{lm}	Cosine coefficient of spherical harmonic potential term		(1.31)
c	Velocity of light	LT^{-1}	(4.35)
E	Eccentric anomaly		(3.13)
E	Elliptic integral of second kind		(3.143)
e	Eccentricity of an ellipse		(3.9)
F	Force	LMT^{-2}	(1.1)
F	Force function per unit mass	L^2T^{-2}	(3.32)
F	Elliptic integral of first kind		(3.135)
F_{lmp}	Inclination function		(3.61)
f	Flattening of ellipse		(1.36)
f	True anomaly		(3.9)
f	Focal length	L	(4.36)
f	Frequency	T^{-1}	(5.5)
$\mathbf{f}$	Vector of residuals		(5.27)
G	Delaunay orbital element, angular momentum per unit mass or $[\mu a(1 - e^2)]^{1/2}$	L^2T^{-1}	(3.41)
G_{lpq}	Eccentricity function		(3.66)

Symbol	Definition	Dimension	First Used in Equation
g_e	Acceleration of gravity at equator	LT^{-2}	(1.38)
H	Delaunay orbital element, $[\mu a(1 - e^2)]^{1/2} \cos i$	L^2T^{-1}	(3.41)
h	Angular momentum per unit mass $r^2\dot{f}$ or $[\mu a(1 - e^2)]^{1/2}$	L^2T^{-1}	(3.4)
h	Delaunay element, node Ω or (node-*GST*): $(\Omega - \theta)$		(3.138)
h	Altitude	L	(4.50)
$\mathbf{I}$	Identity matrix		(2.20)
i	Inclination		(2.29)
J_n	Coefficient of zonal potential term		(1.38)
J_{22}	Coefficient of equatorial ellipticity potential term		(6.10)
$\mathbf{K}$	Covariance matrix of linear transform of observations, $\mathbf{C}\,\mathbf{W}\,\mathbf{C}^T$		(5.45)
K_{ilmpq}	Coefficient of sinusoidal partial derivative with respect to potential coefficient		(4.27)
k	Gravitational constant 6.670×10^{-8} cm^3 gm^{-1} sec^{-2}	$L^3M^{-1}T^{-2}$	(1.1)
k	Modulus of elliptic integral		(3.134)
L	Delaunay orbital element, $\mu^{1/2}a^{1/2}$	L^2T^{-1}	(3.41)
l	Degree subscript of spherical harmonic		(1.16)
$\mathbf{l}_T$	Vector referring to local vertical	L	(4.11)
M	Mass of earth	M	(1.1)
M	Mean anomaly		(3.19)
$\mathbf{M}$	Matrix of coefficients of corrections to parameters		(5.27)
m	Mass of small body	M	(1.1)
m	Order subscript of spherical harmonic		(1.19)
m	Ratio of centrifugal acceleration to equatorial gravity		(1.38)
$\mathbf{N}$	Matrix of normal equation coefficients		(5.53)
n	Mean motion	T^{-1}	(3.20)
n	Wave number in time series		(5.1)
$\mathbf{O}$	Null matrix		(5.43)
O_i	Quantity observed		(4.32)
$\mathbf{P}$	Error propagation matrix		(5.63)
P_{lm}	Legendre associated polynomial		(1.30)
P_j	Parameter		(4.32)
p	Inclination function subscript		(3.61)
$\mathbf{p}$	Vector of momentum (per unit mass) variables	L^2T^{-1} or LT^{-1}	(3.83)
p_i	Momentum (per unit mass) variable	L^2T^{-1} or LT^{-1}	(3.83)

Symbol	*Definition*	*Dimension*	*First Used in Equation*
$\mathbf{p}_T$	Vector referred to inertially fixed camera axis	L	(4.16)
Q_{lm}	Amplitude of term in spherical harmonic potential disturbing function		(3.121)
q	Eccentricity function subscript		(3.67)
$\mathbf{q}$	Position vector referred to orbital plane	L	(3.23)
$\mathbf{q}$	Vector of angle or position variables	0 or L	(3.83)
q_i	Angle or position variable	0 or L	(3.83)
R	Radial function in spherical harmonic		(1.14)
R	Disturbing function	L^2T^{-2}	(3.40)
R_E	Mean radius of the earth	L	(4.68)
$\mathbf{R}_i(\theta)$	Matrix for rotation about axis i through angle θ		(2.6)
$\mathbf{R}_{xq}$	Matrix for rotation from $\boldsymbol{q}$ to $\boldsymbol{x}$ coordinates		(2.26)
r	Radial coordinate or distance	L	(1.1)
$\mathbf{r}, \dot{\mathbf{r}}, \ddot{\mathbf{r}}$	Position, velocity, and acceleration vectors	L, LT^{-1}, LT^{-2}	(3.1)
S	Determining function of canonical transformation		(3.96)
S_{lm}	Sine coefficient of spherical harmonic potential term		(1.31)
S_{lmi}	Surface spherical harmonic		(1.31)
$\mathbf{s}$	Vector of normal equation constants		(5.53)
s_k	Keplerian element		(3.27)
T	Kinetic energy per unit mass	L^2T^{-2}	(3.32)
T	Duration of observation of time series	T	(5.1)
T_{lmt}	Coefficient of term in series development of spherical harmonic		(1.27)
t	Time	T	(3.4)
U	Potential of gravity of reference ellipsoid	L^2T^{-2}	(1.38)
u	Reciprocal $1/r$ of radius	L^{-1}	(3.5)
$\mathbf{u}$	Vector of position referred to earth-fixed coordinates	L	(2.24)
V	Gravitational potential (per unit mass)	L^2T^{-2}	(1.3)
$\mathbf{V}$	Covariance matrix of parameters		(5.50)
V_0	Gravitational potential of reference ellipsoid	L^2T^{-2}	(1.38)
v	Velocity	LT^{-1}	(3.34)
W	Potential of gravity	L^2T^{-2}	(1.35)
$\mathbf{W}$	Covariance matrix of observations		(5.38)
$\mathbf{x}$	Vector of position referred to inertially fixed coordinates	L	(2.24)
$\mathbf{x}$	Vector of corrections to observations		(5.27)
$\mathbf{y}$	Vector time series		(5.1)

Symbol	*Definition*	*Dimension*	*First Used in Equation*
$\{x, y\}$	Plate coordinates	L	(4.36)
z	Zenith angle		(4.18)
$\mathbf{z}$	Vector of corrections to parameters		(5.27)
α	Right ascension		(3.54)
δ	Declination		(4.16)
ϵ	Error		(4.34)
θ	Greenwich sidereal time		(2.24)
$\varkappa$	Swing angle of camera		(4.18)
Λ	Longitude function in spherical harmonic		(1.14)
λ	Longitude		(1.9)
$\boldsymbol{\lambda}$	Vector of Lagrangian multipliers		(5.42)
λ_{lm}	Reference longitude of term in spherical harmonic potential		(3.122)
μ	Sine of latitude, sin ϕ		(1.21)
μ	Gravitational constant times earth's mass kM	L^3T^{-2}	(3.1)
μ	Index of refraction		(4.50)
ν	Radius of curvature in prime vertical	L	(4.8)
ρ	Density	ML^{-3}	(1.5)
σ^2	Degree variance of spherical harmonic		(5.13)
Φ	Latitude function in spherical harmonic		(1.14)
ϕ	Latitude		(1.9)
ψ	Longitude measured from minor axis of equatorial ellipticity		(3.130)
Ω	Longitude of node		(2.28)
ω	Rotation rate $\dot{\theta}$ of earth	T^{-1}	(1.35)
ω	Argument of perigee		(2.30)

TABLE OF NUMERICAL VALUES

Symbol	*Name*	*Value*
A/m	Area-to-mass ratio of satellite, usually	0.02 to 0.20 cm^2/gm
	except balloons	30 to 100 cm^2/gm
a_e	Semimajor axis of earth	6378153 $\pm$ 8 m
b	Semiminor axis of earth	6356768 $\pm$ 8 m
C_D	Satellite shape factor	2.4 $\pm$ 0.2
C_{lm}	Potential coefficient	See Table 3
c	Velocity of light	299792.5 $\pm$ 0.5 km/sec
f	Flattening of earth	1/298.25 $\pm$ 0.01
g_e	Equatorial gravity	978.0284 $\pm$ 0.0013 cm/sec^2
J_l	Zonal potential coefficient	See Table 4
J_{22}	Equatorial ellipticity coefficient	1.8 $\pm$ 0.1 $\times 10^{-6}$
k	Gravitational constant	6.670 $\pm$ 0.002 $\times 10^{-8}$ $cm^3/gm/sec^2$
kM	Gravitational constant $\times$ earth's mass	3.986009 $\pm$ 0.000010 $\times$ $\times 10^{14}$ m^3/sec^2
m	Centrifugal acceleration/equatorial gravity	0.0034678
R_E	Mean radius of earth	6371.0 km
S_{lm}	Potential coefficient	See Table 3
t	Time: in planetary units ($k = 1$, $M = 1$, $a_e = 1$), time unit is	806.8137 secs
v	Velocity: for close satellites	6 to 8.5 km/sec
λ_{22}	Longitude of major axis, equatorial ellipticity	14.5° $\pm$ 1.5° W
μ	kM	3.986009 $\times 10^{14}$ m^3/sec^2
μ	Refractive index	< 1.000390

Symbol	*Name*	*Value*
ρ	Atmospheric density, at altitude:	500 km: 15×10^{-17} to 4×10^{-14} gm/cm^3 1000 km: 2×10^{-19} to 10^{-16} gm/cm^3 1500 km: 1.5×10^{-20} to 8×10^{-18} gm/cm^3
σ_l^2	Degree variance of potential	$\sim 160 \times 10^{-12}/l^3$
$\omega = \dot{\theta}$	Rate of earth's rotation	$0.7292115085 \times 10^{-4}$ sec^{-1}

Theory of Satellite Geodesy

1

THE EARTH'S GRAVITATIONAL FIELD

1.1. Potential Theory

The most pervasive fact of the familiar physical environment is the earth's gravitational attraction. According to Newton's universal law of gravitation, the force of attraction between two particles of masses m and M at a distance r from each other will be

$$F = k\frac{mM}{r^2}, \tag{1.1}$$

where k is the gravitational constant. If we combine Equation (1.1) with Newton's second law, which is

$$F = ma,$$

we obtain the acceleration of the particle of mass m with respect to the center of mass of the two particles,

$$a = \frac{kM}{r^2}. \tag{1.2}$$

This acceleration is the magnitude of a vector directed along the line between the two particles. A vector **a** equivalent to Equation (1.2) will be obtained by expressing the acceleration as the gradient of a scalar, called a potential. Thus

$$\mathbf{a} = \nabla V, \tag{1.3}$$

where

$$V = \frac{kM}{r}. \tag{1.4}$$

In (1.4), V is shown as a positive quantity, which is consistent with the sign convention of astronomy and geodesy. In physics V is conventionally taken to be negative.

For m negligibly small compared to M, Equations (1.3) and (1.4) are consistent with a coordinate system whose origin is at the center of mass of the particle of mass M. For the effect of several particles of masses M_i at distances r_i, the combined acceleration can be expressed as the gradient of a potential, which is a sum of potentials V_i expressed by Equation (1.4). If these particles are conglomerated to form a continuous body of variable density ρ, this summation can be replaced by an integration over the volume of the body. Thus

$$V = k \iiint \frac{\rho(x, y, z)}{r(x, y, z)} \, dx \, dy \, dz. \tag{1.5}$$

For a particular component a_x of $\mathbf{a}$ derived from the point mass potential of Equation (1.4), we have

$$a_x = \frac{\partial V}{\partial x} = -kM \frac{x}{r^3}, \tag{1.6}$$

and for the second derivative, we have

$$\frac{\partial^2 V}{\partial x^2} = kM\left(-\frac{1}{r^3} + \frac{3x^2}{r^5}\right). \tag{1.7}$$

Adding together the second derivatives for the other two coordinates, we get Laplace's equation,

$$\nabla^2 V = \frac{\partial^2 V}{\partial x^2} + \frac{\partial^2 V}{\partial y^2} + \frac{\partial^2 V}{\partial z^2} = kM\left(-\frac{3}{r^3} + \frac{3(x^2 + y^2 + z^2)}{r^5}\right) = 0. \tag{1.8}$$

We would get this same result for any element of mass $\rho \, dx \, dy \, dz$ in the potential of Equation (1.5) and hence for the summation thereof.

The coordinate system, rectangular or otherwise, that is most convenient in a physical problem usually depends on the geometry of the boundaries. The earth is rather round, which suggests spherical coordinates. Thus,

$$\begin{aligned} x &= r \cos \phi \cos \lambda, \\ y &= r \cos \phi \sin \lambda, \\ z &= r \sin \phi, \end{aligned} \tag{1.9}$$

where r is the radial distance from the origin, ϕ is latitude, and λ is longitude measured eastward (that is, counterclockwise looking toward the origin from the positive end of the z-axis). The notation used in Equation (1.9) is consistent with geodetic practice; usually in mathematics there are used θ for colatitude and ϕ for longitude.

In order to convert the Laplace equation, (1.8), to spherical coordinates, we need partial derivatives of the spherical coordinates with respect to the rectangular coordinates. These can be obtained by differentiating Equation (1.9) with respect to r, ϕ, and λ in turn and then by solving the simultaneous differential equations for dr, $d\phi$, $d\lambda$. Thus we have

$$\begin{aligned} dr &= \cos\phi\cos\lambda\,dx + \cos\phi\sin\lambda\,dy + \sin\phi\,dz, \\ d\phi &= -\frac{1}{r}\sin\phi\cos\lambda\,dx - \frac{1}{r}\sin\phi\sin\lambda\,dy + \frac{1}{r}\cos\phi\,dz, \\ d\lambda &= -\frac{1}{r\cos\phi}\sin\lambda\,dx + \frac{1}{r\cos\phi}\cos\lambda\,dy. \end{aligned} \tag{1.10}$$

Then

$$\begin{aligned} \frac{\partial V}{\partial x} &= \frac{\partial V}{\partial r}\frac{\partial r}{\partial x} + \frac{\partial V}{\partial \phi}\frac{\partial \phi}{\partial x} + \frac{\partial V}{\partial \lambda}\frac{\partial \lambda}{\partial x} \\ &= \frac{\partial V}{\partial r}\cos\phi\cos\lambda - \frac{\partial V}{\partial \phi}\frac{1}{r}\sin\phi\cos\lambda - \frac{\sin\lambda}{r\cos\phi}\frac{\partial V}{\partial \lambda} \end{aligned} \tag{1.11}$$

and

$$\begin{aligned} \frac{\partial^2 V}{\partial x^2} = &\left[\frac{\partial^2 V}{\partial r^2}\cos\phi\cos\lambda + \left(\frac{\partial V}{\partial \phi}\cdot\frac{1}{r} - \frac{\partial^2 V}{\partial\phi\,\partial r}\right)\frac{1}{r}\sin\phi\cos\lambda\right. \\ &\left. + \frac{\sin\lambda}{r\cos\phi}\left(\frac{\partial V}{\partial\lambda}\frac{1}{r} - \frac{\partial^2 V}{\partial\lambda\,\partial r}\right)\right]\frac{\partial r}{\partial x} + \text{similar factors} \times \frac{\partial\phi}{\partial x} \text{ and } \frac{\partial\lambda}{\partial x}. \end{aligned} \tag{1.12}$$

Carrying out these differentiations for

$$\frac{\partial^2 V}{\partial x^2}, \quad \frac{\partial^2 V}{\partial y^2}, \quad \text{and} \quad \frac{\partial^2 V}{\partial z^2}$$

in turn and adding them together, we obtain for the Laplace Equation in spherical coordinates,

$$r^2\nabla^2 V = \frac{\partial}{\partial r}\left(r^2\frac{\partial V}{\partial r}\right) + \frac{1}{\cos\phi}\frac{\partial}{\partial\phi}\left(\cos\phi\frac{\partial V}{\partial\phi}\right) + \frac{1}{\cos^2\phi}\frac{\partial^2 V}{\partial\lambda^2} = 0. \tag{1.13}$$

1.2 Spherical Harmonics

To express the variations of the potential V in the spherical coordinate system, it would be convenient if V had the form

$$V = R(r)\,\Phi(\phi)\,\Lambda(\lambda). \tag{1.14}$$

By substituting Equation (1.14) in (1.13) and dividing by $R\Phi\Lambda$, we get

$$\frac{1}{R}\frac{d}{dr}\left(r^2\frac{dR}{dr}\right) + \frac{1}{\Phi\cos\phi}\frac{d}{d\phi}\left(\cos\phi\frac{d\Phi}{d\phi}\right) + \frac{1}{\Lambda\cos^2\phi}\frac{d^2\Lambda}{d\lambda^2} = 0. \tag{1.15}$$

Because the first term of Equation (1.15) is the only term that is a function of r, it must be constant—say, as later turns out to be convenient—$l(l+1)$. Carrying out the differentiation and multiplying by R, we have

$$r^2\frac{d^2R}{dr^2} + 2r\frac{dR}{dr} - l(l+1)R = 0. \tag{1.16}$$

The form of Equation (1.16), in which R and each of its derivatives is multiplied by the equivalent power of r, suggests that R is of the form r^k. Substituting this value in Equation (1.16) and solving the resulting equation for k, we get l and $-l-1$ as the two admissible solutions, or

$$R = Ar^l + Br^{-l-1}, \tag{1.17}$$

where A and B are arbitrary constants. In the case of interest to us, a potential in free space vanishing at infinity, A must equal zero.

Substituting from Equation (1.16) into (1.15) and multiplying by $\cos^2\phi$, we obtain a separation of the Λ term. Thus

$$l(l+1)\cos^2\phi + \frac{\cos\phi}{\Phi}\frac{d}{d\phi}\left(\cos\phi\frac{\partial\Phi}{\partial\phi}\right) + \frac{1}{\Lambda}\frac{d^2\Lambda}{d\lambda^2} = 0. \tag{1.18}$$

The last term of Equation (1.18), since it is the only one that is a function of λ, must be constant. On making this constant equal to $-m^2$, we have

$$\Lambda = C\cos m\lambda + S\sin m\lambda, \tag{1.19}$$

where C and S are arbitrary constants. Substituting $-m^2$ for Λ term of Equation (1.18) and multiplying by $\Phi/\cos^2\phi$, we get an equation that is solely a function of ϕ. Thus

$$\frac{1}{\cos\phi}\frac{d}{d\phi}\left(\cos\phi\frac{d\Phi}{d\phi}\right)+\left[l(l+1)-\frac{m^2}{\cos^2\phi}\right]\Phi=0 \tag{1.20}$$

or, substituting μ for $\sin\phi$,

$$\frac{d}{d\mu}\left[(1-\mu^2)\frac{d\Phi}{d\mu}\right]+\left[l(l+1)-\frac{m^2}{1-\mu^2}\right]\Phi=0. \tag{1.21}$$

The form of Equation (1.21) with $m = 0$, known as Legendre's equation, is solved by assuming that Φ is represented by a power series in μ (Apostol, 1962, pp. 359–364). In the case of $m \neq 0$, the $1 - \mu^2$ in the denominator makes the simple power series representation inconvenient, and the equation must be solved by cut-and-try. The try that succeeds is to assume that Φ has the form

$$\Phi = (1-\mu^2)^{m/2}\nu(\mu), \tag{1.22}$$

which leads to an equation for ν. Therefore,

$$(1-\mu^2)\frac{d^2\nu}{d\mu^2}-2(m+1)\mu\frac{d\nu}{d\mu}+(l-m)(l+m+1)\nu=0. \tag{1.23}$$

Assuming

$$\nu=\sum_{i=0}^{\infty}a_i\mu^i, \tag{1.24}$$

substituting in Equation (1.23), and requiring the coefficient of each power of μ to be separately zero, we obtain a recurrence relationship between alternate coefficients of the power series. From the coefficients of μ^k we get

$$a_{k+2}=\frac{k(k+2m+1)-(l-m)(l+m+1)}{(k+1)(k+2)}a_k. \tag{1.25}$$

In order to obtain the maximum possible value of k, we set the numerator of Equation (1.25) equal to zero and solve for k. The result is

$$k_{\max}=l-m. \tag{1.26}$$

Hence $m \leqq l$, and if ν is to be represented by a finite power series in μ, the allowable powers will be $l - m - 2t$, where t is any non-negative integer $\leqq (l - m)/2$. Let us substitute $(l - m - 2t)$ for k and the notation T_{lmt} for a_k. Then, on taking advantage of the cancellation between some terms in the numerator to obtain a more compact expression, Equation (1.25) becomes

$$T_{lmt} = -\frac{(l - m - 2t + 1)(l - m - 2t + 2)}{2t(2l - 2t + 1)} T_{lm(t-1)}. \qquad (1.27)$$

This solution still leaves T_{lm0} to be defined. Because the whole expression is multiplied by the arbitrary constants BC and BS from Equations (1.17) and (1.19), T_{lm0} is also arbitrary; any change in the value adopted for T_{lm0} would merely result in an inversely proportionate change in BC and BS. As a result of the most common manner of derivation, T_{lm0} is usually defined as (see, for example, Hobson, 1961, p. 91)

$$T_{lm0} = \frac{(2l)!}{2^l l!\,(l - m)!}. \qquad (1.28)$$

Applying Equation (1.27) successively to (1.28), we then get

$$T_{lmt} = \frac{(-1)^t(2l - 2t)!}{2^l t!\,(l - t)!\,(l - m - 2t)!}. \qquad (1.29)$$

The solution Φ of Equation (1.20) or (1.21) corresponding to a particular pair of subscripts l, m is called a Legendre associated function, $P_{lm}(\sin\phi)$ or $P_{lm}(\mu)$; thus we have

$$P_{lm}(\sin\phi) = \cos^m\phi \sum_{t=0}^{k} T_{lmt} \sin^{l-m-2t}\phi, \qquad (1.30)$$

where k is the integer part of $(l - m)/2$.

The complete real solution of the Laplace equation, (1.13), from (1.17) (setting $A = 0$, $B = 1$), (1.19), and (1.30) is then

$$V = \sum_{lmi} \frac{1}{r^{l+1}} S_{lmi} = \sum_{l=0}^{\infty} \sum_{m=0}^{l} \frac{1}{r^{l+1}} P_{lm}(\sin\phi)[C_{lm}\cos m\lambda + S_{lm}\sin m\lambda], \qquad (1.31)$$

where the i subscript in the first term denotes the $\cos m\lambda$ or $\sin m\lambda$ term.

In addition to these real solutions, there are imaginary solutions that are not applicable to the potential problem in which we are interested.

An important property of the surface spherical harmonics S_{lmi} is that they are orthogonal; namely,

$$\int_{\text{sphere}} S_{lmi} S_{hkj}\, d\sigma = 0 \quad \text{if } l \neq h \quad \text{or} \quad m \neq k \quad \text{or} \quad i \neq j \qquad (1.32)$$

for integration over the surface of a sphere. This property makes the spherical harmonics the natural means for general representation of a function over a spherical surface, analogous to Fourier series for a function in a rectilinear space. The integral of the square of S_{lmi} is, for unit C_{lm} or S_{lm},

$$\int_{\text{sphere}} S_{lmi}^2 \, d\sigma = \left[\frac{(l+m)!}{(l-m)!\,(2l+1)(2-\delta_{0m})}\right] 4\pi, \tag{1.33}$$

where the Kronecker delta δ_{0m} is equal to 1 for $m = 0$ and 0 for $m \neq 0$. The factor $(l+m)!/(l-m)!$ indicates that the magnitude of the functions

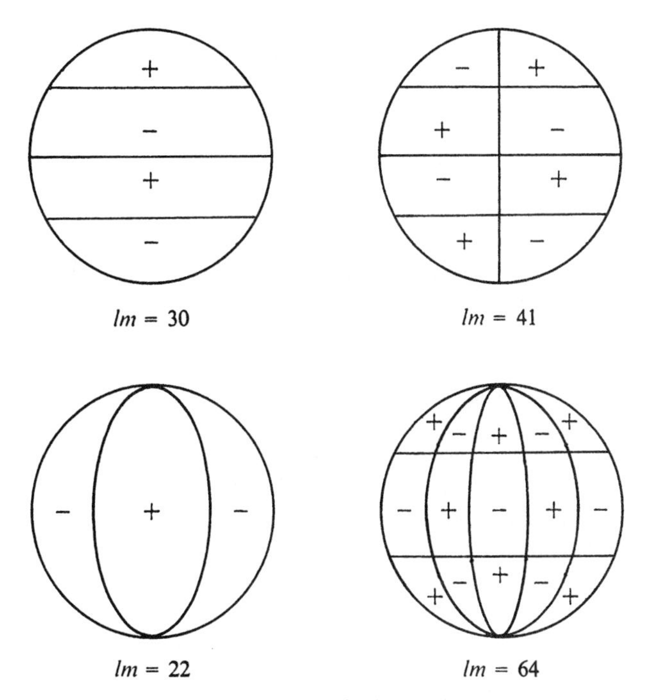

FIGURE 1. *Examples of spherical harmonics.*

(and hence of the coefficients) will vary greatly with the subscript m. In order to make coefficients more readily comparable in numerical work, it is generally convenient to use normalized functions, for example,

$$\bar{S}_{lmi} = \left[\frac{(l-m)!\,(2l+1)(2-\delta_{0m})}{(l+m)!}\right]^{1/2} S_{lmi}. \tag{1.34}$$

Spherical harmonics are most conveniently remembered in terms of their zeros. A surface harmonic S_{lmi} will have $(l-m)$ zeros in a distance π along a meridian and m zeros in the same distance along a parallel. Some examples are shown in Figure 1.

1.3 Potential of the Ellipsoid

The shape of the earth, as defined very closely by the geoid, or mean sea level (less meteorological effects), is determined not only by the gravitational potential V but also by the potential of rotation. These two potentials combine to make what is called the potential of gravity,

$$W = V + \tfrac{1}{2}\omega^2 r^2 \cos^2 \phi, \tag{1.35}$$

where ω is the rate of rotation.

It is an observed fact that the geoid is approximated within about 10^{-5} of the radius vector by an ellipsoid of revolution. The shape of this ellipsoid is conventionally expressed by the flattening f,

$$f = (a - b)/a, \tag{1.36}$$

where a is the equatorial radius and b is the polar radius. The value of f is about 3.353×10^{-3}. Hence, to explore the 10^{-5} departures of the earth's gravitational field from a reference potential V_0 of an ellipsoid, it is necessary to carry the expression of V_0 in terms of the mass M, radius a, rotation ω, and flattening f to terms of order f^2. Furthermore, in order to connect results of systems affected by gravitation (satellite orbits) with those affected by gravity (gravimetry), it is desirable to connect the parameters of the ellipsoid convenient to the former (kM and C_{20}) to those convenient to the latter (the equatorial acceleration of gravity g_e and f). The algebra involved in making this connection is considerable, so we shall write out only the terms to $O(f)$ in the solution outlined here.

The symmetries of an ellipsoid of revolution indicate that its radius vector can be expressed as a sum of even degree zonal harmonics. Thus, we have

$$r = r_0(1 + \alpha_2 P_{20} + \alpha_4 P_{40} + \cdots), \tag{1.37}$$

where P_{20}, P_{40} are defined by (1.29) and (1.30).

The customary manner of representing the potential of gravity of a reference figure, or "normal" potential, is

$$U = \frac{kM}{r}\left[1 - J_2\left(\frac{a}{r}\right)^2 P_{20} - J_4\left(\frac{a}{r}\right)^4 P_{40} - \cdots\right] + \frac{1}{3}\frac{g_e m}{a} r^2(1 - P_{20}), \tag{1.38}$$

where

$$m = \frac{\omega^2 a}{g_e}, \tag{1.39}$$

in which g_e is the acceleration of gravity at the equator.

Expanding (1.37) by the binomial theorem gives

$$r^n = r_0^n\left\{1 + \frac{n(n-1)}{10}\alpha_2^2 + \left[n\alpha_2 + \frac{n(n+1)}{7}\alpha_2^2\right]P_{20}\right.$$

$$\left. + \left[n\alpha_4 + \frac{9n(n-1)}{35}\alpha_2^2\right]P_{40} + \cdots\right\}, \tag{1.40}$$

when use has been made of

$$P_{20}^2 = \tfrac{18}{35}P_{40} + \tfrac{2}{7}P_{20} + \tfrac{1}{5}. \tag{1.41}$$

Substituting (1.40) into (1.38), using (1.41), and neglecting $P_{20}P_{40}$ and P_{40}^2 terms, we get a form for the potential U on the surface of the ellipsoid. Thus

$$\begin{aligned} U = {} & U_0(g_e, m, r_0, \alpha_2, a, kM, J_2) \\ & + C_2(g_e, m, r_0, \alpha_2, a, kM, J_2)P_{20} \\ & + C_4(g_e, m, r_0, \alpha_2, \alpha_4, a, kM, J_2, J_4)P_{40}, \end{aligned} \tag{1.42}$$

where

$$U_0 = \frac{kM}{r_0} + \frac{g_e m}{3a}r_0^2 + O(f^2),$$

$$C_2 = -\frac{kM}{r_0}\left[\alpha_2 + J_2\left(\frac{a}{r_0}\right)^2\right] - \frac{g_e m}{3a}r_0^2 + O(f^2),$$

$$C_4 = O(f^2).$$

If the ellipsoidal surface is an equipotential, then U must be constant and equal to U_0 thereon. The coefficients C_2 and C_4 must therefore be separately equal to zero, which yields two equations for kM, J_2, and J_4 in terms of the other parameters. A third equation is obtained by the condition that the negative of the radial derivative of U must be equal to g_e at the equator. Thus,

$$-g_e = \frac{-kM}{a^2}\left(1 + \frac{3}{2}J_2 - \frac{15}{8}J_4\right) + g_e m. \tag{1.43}$$

The three equations then can be solved simultaneously for kM, J_2, and J_4. We have

$$kM = a^2 g_e[1 + 3m/2 + 3\alpha_2/2 + O(f^2)],$$

$$J_2 = -m/3 - \alpha_2 + O(f^2),$$

$$J_4 = O(f^2).$$

In order to obtain r_0, α_2, and α_4 in terms of the conventional parameters for an ellipsoid, a and f, we have the equation

$$\frac{x^2}{a^2} + \frac{y^2}{a^2} + \frac{z^2}{b^2} = 1. \tag{1.44}$$

Converting from rectangular to spherical coordinates by Equation (1.9), using (1.36) to eliminate b, solving for r^2, and then applying the binomial theorem to obtain r, we get

$$r = a\left[1 - \left(f + \frac{3}{2}f^2 + \cdots\right)\sin^2\phi + \frac{3}{2}f^2\sin^4\phi - \cdots\right]. \tag{1.45}$$

Integrating (1.45) from 0 to 1 with respect to $\sin\phi$, we get r_0,

$$r_0 = a[1 - f/3 + O(f^2)],$$

and then α_2 and α_4,

$$\alpha_2 = -2f/3 + O(f^2),$$

$$\alpha_4 = O(f^2).$$

Substitution in the equations for kM, J_2, and J_4 gives the final solution,

$$kM = a^2 g_e\left(1 - f + \frac{3}{2}m - \frac{15}{14}mf + \cdots\right), \tag{1.46}$$

$$J_2 = \frac{2}{3}f\left(1 - \frac{1}{2}f\right) - \frac{1}{3}m\left(1 - \frac{3}{2}m - \frac{2}{7}f\right) + \cdots, \tag{1.47}$$

$$J_4 = -\frac{4}{35}f(7f - 5m) + \cdots. \tag{1.48}$$

To summarize, for the purposes of celestial geodesy, we can consider the earth's gravity field as represented by a normal potential of an ellipsoid of revolution, Equation (1.38), plus small irregular variations expressed by a sum of spherical harmonics, as in (1.31).

REFERENCES

1. Apostol, Tom M. *Calculus.* Vol. II. New York: Blaisdell Publishing Company, 1962.

2. Bomford, G. *Geodesy.* 2nd ed. London: Oxford University Press, 1962.

3. Heiskanen, W. A., and F. A. Vening-Meinesz. *The Earth and Its Gravity Field.* New York: McGraw-Hill Book Company, 1958.

4. Hobson, E. W. *The Theory of Spherical and Ellipsoidal Harmonics.* London: Cambridge University Press, 1931.

5. Jeffreys, H. *The Earth.* 4th ed. London: Cambridge University Press, 1959.

6. Jung, K. "Figure der Erde." *Handbuch der Physik*, *47* (1956), pp. 534–639.

7. Kaula, W. M. "Determination of the Earth's Gravitational Field." *Revs. Geophys.*, *1* (1963), pp. 507–552.

8. Kellogg, O. D. *Foundations of Potential Theory.* New York: Dover Publishers Inc., 1953.

2

MATRICES AND ORBITAL GEOMETRY

2.1. General

The purpose of this chapter is to describe the geometry of an idealized situation: a vacuum with an earth rotating uniformly with respect to fixed inertial coordinates. This idealized situation will constitute both the reference frame for the development of close satellite orbit dynamics in Chapter 3, and for the description of observations of satellites and variations in the coordinate system in Chapter 4. As a preliminary to describing the geometry, as well as to some techniques of data analysis in Chapters 5 and 6, we summarize the rules of matrix algebra.

2.2. Matrix Notation

A vector **x**, or $\{x_1, x_2, x_3\}^T$, can be changed into another vector **y** by a linear transformation given by

$$\begin{aligned} y_1 &= a_{11}x_1 + a_{12}x_2 + a_{13}x_3, \\ y_2 &= a_{21}x_1 + a_{22}x_2 + a_{23}x_3, \\ y_3 &= a_{31}x_1 + a_{32}x_2 + a_{33}x_3. \end{aligned} \tag{2.1}$$

Equation (2.1) can be abbreviated as

$$y_i = \sum_j a_{ij}x_j, \qquad i, j = 1,\ 2,\ 3, \tag{2.2}$$

or as

$$\mathbf{y} = \mathbf{A}\mathbf{x}. \tag{2.3}$$

A rectangular array **A** of numbers a_{ij} is called a matrix. Matrix algebra is the expression of algebraic operations on arrays of quantities, such as the transformation in (2.1), in compressed notation such as (2.3). In this text we shall be interested both in expressing transformations from one coordinate system to another and in the formation and solution of generalized least-squares problems. The principal rules of matrix algebra are:

The numbers a_{ij} that comprise a matrix are called *elements*.

The first subscript i denotes the *row*, and the second subscript j denotes the *column*, in accordance with the customary method of displaying matrices. For example, we have

$$\mathbf{A} = [a_{ij}] = \begin{bmatrix} a_{11} & a_{12} & \cdots & a_{1n} \\ a_{21} & a_{22} & \cdots & a_{2n} \\ \cdot & & & \cdot \\ \cdot & & & \cdot \\ \cdot & & & \cdot \\ a_{m1} & a_{m2} & \cdots & a_{mn} \end{bmatrix}. \tag{2.4}$$

The number of rows, m, and the number of columns n are called its *dimensions*. Particular types of matrices are:

1. A *vector*, or column matrix, is a matrix that has only one column. We denote vectors by lower-case boldface letters, such as

$$\mathbf{a} = [a_i] = \begin{bmatrix} a_1 \\ a_2 \\ \cdot \\ \cdot \\ \cdot \\ a_m \end{bmatrix}. \tag{2.5}$$

2. A *square matrix*, or quadratic matrix, has the same number of rows and columns.

3. An *orthogonal matrix* is a square matrix whose determinant $|a_{ij}|$ is ± 1, and whose inverse is equal to its transpose [see (2.11) and (2.20)].

4. A *rotation matrix* is an orthogonal matrix whose determinant is $+1$. In this book we are interested mainly in rotation matrices of dimension 3×3. For those rotation matrices whose elements r_{lm} satisfy the following rules, we adopt the notation $\mathbf{R}_i(\theta)$:

$$\begin{gathered} j \equiv i \,(\text{modulo } 3) + 1, \qquad k \equiv j \,(\text{modulo } 3) + 1, \\ r_{ii} = 1, \qquad r_{ij} = r_{ji} = r_{ik} = r_{ki} = 0, \\ r_{jj} = r_{kk} = +\cos\theta, \qquad r_{jk} = +\sin\theta, \qquad r_{kj} = -\sin\theta. \end{gathered} \tag{2.6}$$

These rules are consistent with a right-handed coordinate system and positive signs for counterclockwise rotation, as viewed looking toward the origin from the positive axis. For example, we have

$$\mathbf{R}_3(\theta) = \begin{bmatrix} \cos\theta & \sin\theta & 0 \\ -\sin\theta & \cos\theta & 0 \\ 0 & 0 & 1 \end{bmatrix}. \tag{2.7}$$

Rotation matrices are also called direction cosine matrices.

5. A *diagonal matrix* is a matrix whose elements a_{ij} satisfy the rule

$$a_{ij} = 0 \qquad \text{if} \quad i \neq j. \tag{2.8}$$

6. An *identity matrix*, or unit matrix, is a diagonal rotation matrix; that is, all a_{ii} are 1. It is generally denoted by **I**.

7. A *null matrix*, or zero matrix, is one all of whose elements are 0. It is generally denoted by **0**.

8. A *symmetric matrix* is one whose elements a_{ij} satisfy the rule

$$a_{ij} = a_{ji}. \tag{2.9}$$

9. An *antisymmetric matrix*, or skew-symmetric matrix, is one whose elements a_{ij} satisfy the rule

$$a_{ij} = -a_{ji}. \tag{2.10}$$

The *transpose* **B** of a matrix **A** is a matrix whose elements b_{ij} satisfy the rule

$$b_{ij} = a_{ji}. \tag{2.11}$$

The transpose of a matrix **A** is generally denoted by $\mathbf{A}^T$.

Operations in matrix algebra are given as follows:

1. The *sum* **C** of two matrices **A** and **B** of equal dimension has elements c_{ij} that satisfy the rule

$$c_{ij} = a_{ij} + b_{ij}. \tag{2.12}$$

The operation of summing is denoted by

$$\mathbf{C} = \mathbf{A} + \mathbf{B}. \tag{2.13}$$

2. The *difference* **D** of **A** and **B** is similarly defined and denoted by

$$\begin{aligned} d_{ij} &= a_{ij} - b_{ij}, \\ \mathbf{D} &= \mathbf{A} - \mathbf{B}. \end{aligned} \tag{2.14}$$

3. The *product* **P** of **A** and **B** has elements p_{ik} that satisfy the rule

$$p_{ik} = \sum_j a_{ij} b_{jk}. \tag{2.15}$$

Hence the number of columns of **A** must equal the number of rows of **B**. In matrix notation, multiplication is denoted by

$$\mathbf{P} = \mathbf{AB}. \tag{2.16}$$

Equations (2.1)–(2.3) express a multiplication in which the matrices **B** and **P** are vectors. Matrix multiplication satisfies the *associative rule*

$$\mathbf{A}(\mathbf{BC}) = (\mathbf{AB})\mathbf{C} \tag{2.17}$$

but, in general, does not satisfy the *commutative rule*

$$\mathbf{AB} \neq \mathbf{BA}, \tag{2.18}$$

also written

$$(\mathbf{AB})^T = \mathbf{B}^T\mathbf{A}^T. \tag{2.19}$$

If the determinant $|a_{ij}|$ of a square matrix **A** is nonzero, then there exists one and only one matrix, which is called the *inverse*, or reciprocal, matrix of **A** and is denoted by $\mathbf{A}^{-1}$, for which

$$\mathbf{AA}^{-1} = \mathbf{A}^{-1}\mathbf{A} = \mathbf{I}. \tag{2.20}$$

The elements of $\mathbf{A}^{-1}$ are given by

$$a_{ji}^{-1} = \frac{K_{ij}}{|a_{ij}|}, \tag{2.21}$$

where K_{ij} is the cofactor of the element a_{ij} in the determinant $|a_{ij}|$, namely, $(-1)^{i+j}$ times the minor obtained from $|a_{ij}|$ by taking away the ith row and the jth column.

The operations of *differentiation* and *integration* of a matrix are applied to each element separately; that is,

$$\frac{\partial \mathbf{A}}{\partial x} = \begin{bmatrix} \partial a_{11}/\partial x & \partial a_{12}/\partial x & \cdots & \partial a_{1n}/\partial x \\ \partial a_{21}/\partial x & \partial a_{22}/\partial x & \cdots & \partial a_{2n}/\partial x \\ \cdot & & & \cdot \\ \cdot & & & \cdot \\ \cdot & & & \cdot \\ \partial a_{m1}/\partial x & \partial a_{m2}/\partial x & \cdots & \partial a_{mn}/\partial x \end{bmatrix} \tag{2.22}$$

and similarly for integration.

A *Jacobian* is a matrix of partial derivatives of the elements of one vector with respect to those of another and is given by

$$\mathbf{J} = \frac{\partial \mathbf{y}}{\partial \mathbf{x}} = \frac{\partial(y_1, y_2, y_3)}{\partial(x_1, x_2, x_3)} = \begin{bmatrix} \partial y_1/\partial x_1 & \partial y_1/\partial x_2 & \partial y_1/\partial x_3 \\ \partial y_2/\partial x_1 & \partial y_2/\partial x_2 & \partial y_2/\partial x_3 \\ \partial y_3/\partial x_1 & \partial y_3/\partial x_2 & \partial y_3/\partial x_3 \end{bmatrix}. \tag{2.23}$$

If the elements of $\mathbf{A}$ in (2.3) are not functions of the elements of $\mathbf{x}$, then $\mathbf{A}$ is the Jacobian of $\mathbf{y}$ with respect to $\mathbf{x}$. If the determinant of $\mathbf{A}$ in (2.3) is 1, then it is a rotation matrix, and (2.3) expresses a rotation of coordinate axes.

2.3. Orbital Geometry

For reasons that will become apparent in Chapter 3, it is convenient to refer the position of a satellite to rectangular coordinates $\mathbf{q}$ fixed in an

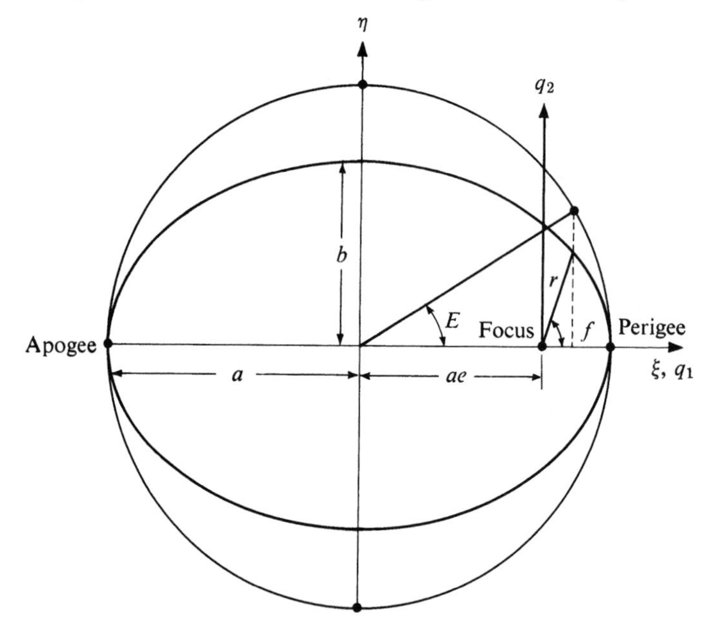

FIGURE 2. *Orbital ellipse.*

ellipse inclined to the equatorial plane, as shown in Figures 2 and 3. Geodesists are interested in earth-fixed coordinates; we need to connect earth-fixed positions to positions referred to this tilted ellipse.

Let earth-fixed positions be represented by a rectangular-coordinate system $\mathbf{u}$, with the u_1 (or u) axis toward latitude 0°, longitude 0°; the u_2

(or v) axis toward latitude 0°, longitude 90°E; and the u_3 (or w) axis toward latitude 90°N, the north pole. The connection is through an inertially fixed-coordinate system $\mathbf{x}$, with the x_1 axis toward the vernal equinox, the point where the sun's orbit intersects the equator; the x_2 axis 90°E eastward in the equator; and the x_3 axis toward the north pole. The angle between the equinox and the Greenwich meridian—0° longitude—is known as the Greenwich Sidereal Time.

Hence, for an earth rotating counterclockwise uniformly about an axis fixed with respect to inertial space, the shift from the earth-fixed coordinate

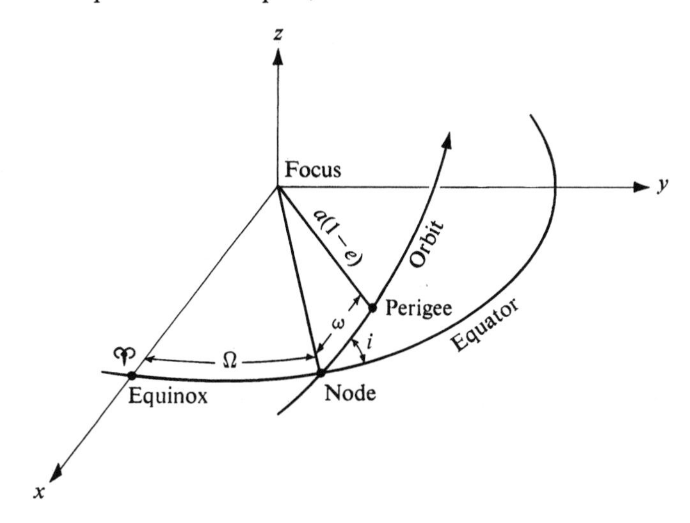

FIGURE 3. *Orbital orientation.*

system $\mathbf{u}$ to the inertially fixed system $\mathbf{x}$ will be a simple clockwise rotation about the W, or Z, axis through θ, the Greenwich Sidereal Time. On following the notation of (2.7),

$$\mathbf{x} = \mathbf{R}_3(-\theta)\mathbf{u} = \begin{bmatrix} \cos\theta & -\sin\theta & 0 \\ \sin\theta & \cos\theta & 0 \\ 0 & 0 & 1 \end{bmatrix} \mathbf{u}. \tag{2.24}$$

Performing the multiplication of (2.24), we have

$$\begin{aligned} x &= u\cos\theta - v\sin\theta, \\ y &= u\sin\theta + v\cos\theta, \\ z &= w. \end{aligned} \tag{2.25}$$

Using the alternative notation of subscripts to the rotation matrix, denoting the vectors transformed by the rotation, we get

$$\mathbf{R}_{xu} = \mathbf{R}_3(-\theta), \tag{2.26}$$

and inversely,

$$\mathbf{R}_{ux} = \mathbf{R}_3(\theta). \tag{2.27}$$

For rotation from the **x** coordinates to the **q** coordinates—with q_1 toward the point of the ellipse closest to the origin (called perigee), q_2 in the orbital plane (as defined by Figure 2), and q_3 normal to the orbital plane—we require first a counterclockwise rotation about the x_3-axis (3-axis) from the vernal equinox to the intersection of the inclined plane with the equator, called the nodes (see Figure 3). This rotation is denoted by

$$\mathbf{R}_3(\Omega) \tag{2.28}$$

Next, a counterclockwise rotation about the 1-axis, from the equatorial plane to the orbital plane is given by

$$\mathbf{R}_1(i)\ \mathbf{R}_3(\Omega). \tag{2.29}$$

And finally a counterclockwise rotation about the 3-axis from the node to perigee is given by

$$\mathbf{R}_{qx} = \mathbf{R}_3(\omega)\ \mathbf{R}_1(i)\ \mathbf{R}_3(\Omega). \tag{2.30}$$

Ω, i, and ω are identical with the Euler angles relating the **q** and **x** coordinate axes. Conversely, we have

$$\mathbf{R}_{xq} = \mathbf{R}_3(-\Omega)\ \mathbf{R}_1(-i)\ \mathbf{R}_3(-\omega). \tag{2.31}$$

Applying (2.6) and multiplying the matrices together, we get

$$\mathbf{R}_{xq} = \begin{bmatrix} \cos\Omega\cos\omega - \sin\Omega\cos i\sin\omega, & -\cos\Omega\sin\omega - \sin\Omega\cos i\cos\omega, & \sin\Omega\sin i \\ \sin\Omega\cos\omega + \cos\Omega\cos i\sin\omega, & -\sin\Omega\sin\omega + \cos\Omega\cos i\cos\omega, & -\cos\Omega\sin i \\ \sin i\sin\omega, & \sin i\cos\omega, & \cos i \end{bmatrix} \tag{2.32}$$

which is required for use in Section 3.2. An alternative notation often used is P, Q, W for the unit vectors along the **q** axes referred to the **x** axes:

$$\mathbf{R}_{xq} = \{P,\ Q,\ W\}. \tag{2.33}$$

REFERENCES

1. Brouwer, D., and G. M. Clemence. *Methods of Celestial Mechanics.* New York: Academic Press, Inc., 1961.

2. Frazer, R. A., W. I. Duncan, and A. R. Collar. *Elementary Matrices.* London: Cambridge University Press, 1938.

3. Goldstein, H. *Classical Mechanics.* Reading, Massachusetts: Addison-Wesley Publishing Company, Inc., 1950.

4. Hawkins, G. A. *Multilinear Analysis for Students in Engineering and Science.* New York: John Wiley and Sons, Inc., 1963.

5. Kaula, W. M. "Analysis of Gravitational and Geometric Aspects of Geodetic Utilization of Satellites." *Geophys. J.*, *5* (1961), pp. 104–133.

6. Veis, G. "Geodetic Uses of Satellites." *Smithsonian Contrib. to Astrophys.*, *3* (1960), pp. 95–161.

3

SATELLITE ORBIT DYNAMICS

3.1. Elliptic Motion

Let us assume that we have a particle of negligible mass attracted by another point mass M in accordance with Equation (1.2). Let us assume further that the origin of coordinates is at the mass M. Equation (1.2) for the acceleration of the particle can then be expressed in vectorial form, using μ for kM, as

$$\ddot{\mathbf{r}} = -kM\mathbf{r}/r^3 = -\mu\mathbf{r}/r^3. \tag{3.1}$$

The acceleration vector $\ddot{\mathbf{r}}$ is therefore colinear with the position vector $\mathbf{r}$. If we define the equatorial plane as the plane determined by the position vector and the velocity vector $\dot{\mathbf{r}}$, the particle will never depart from the equatorial plane because there is no component of acceleration out of the plane. Hence, in converting from rectangular to spherical coordinates by Equation (1.9), we can set equal to zero the latitude ϕ and its derivatives with respect to time $\dot{\phi}$ and $\ddot{\phi}$. Differentiating Equation (1.9) twice with respect to time, we obtain Equations (3.1) in polar coordinates,

$$\begin{aligned}
x &= r\cos\lambda,\\
y &= r\sin\lambda,\\
\dot{x} &= \dot{r}\cos\lambda - r\dot{\lambda}\sin\lambda,\\
\dot{y} &= \dot{r}\sin\lambda + r\dot{\lambda}\cos\lambda,\\
\ddot{x} &= \ddot{r}\cos\lambda - 2\dot{r}\dot{\lambda}\sin\lambda - r\ddot{\lambda}\sin\lambda - r(\dot{\lambda})^2\cos\lambda = -\mu\cos\lambda/r^2,\\
\ddot{y} &= \ddot{r}\sin\lambda + 2\dot{r}\dot{\lambda}\cos\lambda + r\ddot{\lambda}\cos\lambda - r(\dot{\lambda})^2\sin\lambda = -\mu\sin\lambda/r^2.
\end{aligned}$$

The point from which the longitude λ is measured is arbitrary, so we can also set λ = zero, but not $\dot{\lambda}$ or $\ddot{\lambda}$. The equations of motion thus become

$$\ddot{r} - r(\dot{\lambda})^2 = -\mu/r^2, \tag{3.2}$$

$$r\ddot{\lambda} + 2\dot{r}\dot{\lambda} = 0. \tag{3.3}$$

If we multiply (3.3) by r, it is evident that the equation is immediately integrable to

$$r^2\dot{\lambda} = h, \tag{3.4}$$

where h is constant. Equation (3.4) states that angular momentum, $r^2\dot{\lambda}$, is conserved. We integrate (3.2), replacing $1/r$ by u. Then

$$\frac{du}{dr} = -\frac{1}{r^2}.$$

From (3.4),

$$\frac{dt}{d\lambda} = \frac{r^2}{h},$$

whence

$$\frac{du}{d\lambda} = \frac{du}{dr}\cdot\frac{dr}{dt}\cdot\frac{dt}{d\lambda} = -\frac{1}{r^2}\dot{r}\frac{r^2}{h} = -\frac{\dot{r}}{h}.$$

Also

$$\frac{d^2u}{d\lambda^2} = \frac{d}{dt}\left(-\frac{\dot{r}}{h}\right)\frac{dt}{d\lambda} = -\frac{\ddot{r}}{h}\frac{r^2}{h} = -\frac{\ddot{r}}{u^2h^2}$$

or

$$\ddot{r} = -h^2u^2\frac{d^2u}{d\lambda^2}. \tag{3.5}$$

Substituting from (3.4) for $\dot{\lambda}$, and from (3.5) for $\ddot{r}$ in (3.2), and replacing r by $1/u$ everywhere, we get

$$\frac{d^2u}{d\lambda^2} + u = \frac{\mu}{h^2}. \tag{3.6}$$

Equation (3.6) is readily integrated as

$$\frac{1}{r} = u = A\cos(\lambda - \lambda_0) + \frac{\mu}{h^2}. \tag{3.7}$$

If in the equation for an ellipse (see Figure 2), with origin at the center,

$$\frac{\xi^2}{a^2} + \frac{\eta^2}{b^2} = 1, \tag{3.8}$$

we substitute $ae + r\cos f$ for ξ, $r\sin f$ for η, and $a^2(1 - e^2)$ for b^2, and solve the resulting quadratic equation for r, we get for the positive root

$$r = \frac{a(1 - e^2)}{(1 + e\cos f)}$$

or

$$\frac{1}{r} = \frac{1}{a(1 - e^2)} + \frac{e}{a(1 - e^2)}\cos f. \tag{3.9}$$

Comparing (3.7) and (3.9), we see that (3.7) is the equation of an ellipse with origin at the focus and that

$$\lambda - \lambda_0 = f, \tag{3.10}$$

$$A = \frac{e}{a(1 - e^2)}, \tag{3.11}$$

$$h = \sqrt{\mu a(1 - e^2)}. \tag{3.12}$$

The size of the orbit of the particle can thus be expressed by the semimajor axis a of the ellipse; the shape, by the eccentricity e; and the location of the particle in the ellipse by f, called the true anomaly. Position in the orbital plane can also be expressed by the **q**-coordinate system, described in Section 2.3 and shown in Figure 2. In order to specify completely the location of the particle, we need the three Euler angles shown in Figure 3 and described as rotations in (2.28)–(2.30): the longitude of the node Ω, the inclination i, and the argument of perigee ω.

Another way of locating the particle in the ellipse, which is sometimes more convenient, is the eccentric anomaly E. The eccentric anomaly, as shown in Figure 2, is the angle subtended at the center of a circle of radius a tangent to the ellipse by the point on the circle whose ξ coordinate is the same as that of the point on the ellipse. From Figure 2 we get

$$q_1 = \xi - ae = a(\cos E - e). \tag{3.13}$$

Using (3.8), we then obtain

$$q_2 = \eta = a\sqrt{1 - e^2}\sin E, \tag{3.14}$$

$$r = \sqrt{q_1^2 + q_2^2} = a(1 - e\cos E). \tag{3.15}$$

For the rate of motion of the particle in its orbit, we can use (3.4), changing $\dot{\lambda}$ to $\dot{f}$. Equation (3.4) is more readily integrated if the true anomaly is replaced by the eccentric anomaly. Differentiating (3.9) with respect to f, we get

$$\frac{dr}{df} = -\frac{r^2\,d(1/r)}{df} = \frac{r^2 e}{a(1-e^2)}\sin f.$$

If we substitute q_2 for $r\sin f$ from Figure 2,

$$dr = \frac{req_2}{a(1-e^2)}\,df. \tag{3.16}$$

From differentiating (3.15) and (3.14), we obtain

$$dr = \frac{e}{\sqrt{1-e^2}}\,q_2\,dE. \tag{3.17}$$

On using (3.16) and (3.17) to eliminate df, (3.15) for r, and (3.12) for h, (3.4) becomes

$$a^2\sqrt{1-e^2}\,(1-e\cos E)\,dE = \sqrt{\mu a(1-e^2)}\,dt. \tag{3.18}$$

Equation (3.18) integrates to

$$E - e\sin E = M, \tag{3.19}$$

where

$$M = n(t - t_0)$$

and

$$n = \mu^{1/2}a^{-3/2}. \tag{3.20}$$

The time t_0 is the time of passing perigee. The quantity M is known as the mean anomaly, and the quantity n as the mean motion. Equation (3.19) is known as Kepler's equation and (3.20) as Kepler's third law.

The angular momentum per unit mass h, whose magnitude is given by (3.4) or (3.12), can also be expressed as the vector cross product of the position and velocity,

$$\mathbf{h} = \mathbf{x} \times \dot{\mathbf{x}}. \tag{3.21}$$

On using the rotation matrix $\mathbf{R}_{xq}$ defined by (2.32), the inertial rectangular coordinates can be expressed in terms of the Keplerian elements,

$$\begin{aligned}\mathbf{x} &= \mathbf{R}_{xq}\{\Omega,\ i,\ \omega\}\,\mathbf{q}\{a,\ e,\ M\},\\ \dot{\mathbf{x}} &= \mathbf{R}_{xq}\{\Omega,\ i,\ \omega\}\,\dot{\mathbf{q}}\{a,\ e,\ M\},\end{aligned} \tag{3.22}$$

where, from (3.13), (3.14) and Figure 2,

$$\mathbf{q} = \begin{bmatrix} a(\cos E - e) \\ a\sqrt{1 - e^2}\sin E \\ 0 \end{bmatrix} = \begin{bmatrix} r\cos f \\ r\sin f \\ 0 \end{bmatrix}; \tag{3.23}$$

and from (3.13)–(3.18),

$$\dot{\mathbf{q}} = \begin{bmatrix} -\sin E \\ \sqrt{1 - e^2}\cos E \\ 0 \end{bmatrix} \frac{na}{1 - e\cos E} = \begin{bmatrix} -\sin f \\ e + \cos f \\ 0 \end{bmatrix} \frac{na}{\sqrt{1 - e^2}}. \tag{3.24}$$

For the velocity v, using (3.9) with (3.20) and (3.24), we have

$$\begin{aligned} v^2 &= \dot{q}_1^2 + \dot{q}_2^2 \\ &= \frac{n^2a^2}{(1 - e^2)}(\sin^2 f + e^2 + 2e\cos f + \cos^2 f) \\ &= \frac{\mu}{a(1 - e^2)}[(2 + 2e\cos f) - (1 - e^2)] \\ &= \mu\left(\frac{2}{r} - \frac{1}{a}\right). \end{aligned}$$

Then for the total energy per unit mass, following the sign convention of physics, we have

$$T - V = \frac{v^2}{2} - \frac{\mu}{r} = -\frac{\mu}{2a}. \tag{3.25}$$

In order to perform the reverse of (3.22), that is, to go from rectangular components to Keplerian elements, the fact that the angular momentum vector $\mathbf{h}$ of (3.21) is normal to the orbital plane can be used to determine the longitude of the node Ω and the inclination i. Referring to Figure 3, we see that

$$\begin{aligned} \Omega &= \tan^{-1}[h_1/(-h_2)], \\ i &= \tan^{-1}[(h_1^2 + h_2^2)^{1/2}/h_3], \end{aligned}$$

where h_1, h_2, h_3 are the components of $\mathbf{h}$. Then in the orbital plane let

$$\mathbf{p} = \mathbf{R}_1(i)\ \mathbf{R}_3(\Omega)\mathbf{x},$$

whence

$$\omega + f = \tan^{-1}(p_2/p_1).$$

Also needed

$$\dot{r} = \mathbf{x} \cdot \dot{\mathbf{x}}/r;$$

from (3.25),

$$a = r\mu/(2\mu - rv^2);$$

from (3.12),

$$e = (1 - h^2/\mu a)^{1/2}.$$

Then from (3.15), (3.17), and (3.18), we have

$$\cos E = (a - r)/(ae),$$

$$\sin E = r\dot{r}/e(\mu a)^{1/2},$$

and finally from (3.23),

$$f = \tan^{-1}[\sqrt{1 - e^2}\sin E/(\cos E - e)].$$

3.2. Perturbed Equations of Motion

The foregoing developments apply solely to motion in a purely central field, but our interest in satellite geodesy is mainly due to the fact that the earth's gravitational field is noncentral; that is, Equation (3.1) should be replaced by

$$\ddot{\mathbf{r}} = \nabla V,$$

where V has a noncentral form such as (1.31) or (1.38). However, even for this noncentral field the Keplerian ellipse and its orientation can be regarded as a coordinate system, alternative to rectangular or polar coordinates, analogous to the use of geodetic latitude and longitude and altitude for position in an earth-fixed system. At any instant the situation of a satellite in earth-centered, inertially fixed coordinates can be described by the rectangular components of position $\{x, y, z\}$ and velocity $\{\dot{x}, \dot{y}, \dot{z}\}$. In place of these six numbers the six numbers of the Keplerian ellipse $\{a, e, i, M, \omega, \Omega\}$ may be used. The relationship between the two systems can be expressed by the rotation from a coordinate system in the orbital plane referred to perigee to the inertially fixed system, as given by (3.22).

The Keplerian ellipse $\{a, e, i, M, \omega, \Omega\}$ corresponding to the position $\mathbf{r}$ and velocity $\dot{\mathbf{r}}$ of a particle at a particular time is known as the instantaneous, or osculating, orbit. If the potential field V differs from a central field, this ellipse will be continually changing. However, if the field differs very slightly from a central field—as is the case for the earth—we should expect that the parameters of the ellipse would change slowly, and hence that the ellipse would constitute a coordinate system convenient for representing the position and velocity of the particle. The problem is to convert the equations of motion from rectangular coordinates to Keplerian ellipse coordinates, or elements, as they are more conventionally called. First we convert from vectorial to subscript notation, and second we change the equations of motion from three second-order equations to six first-order equations by treating the velocity components as variables the same as the position components. Accordingly,

$$\begin{aligned} \frac{d}{dt} x_i &= \dot{x}_i, \qquad i = 1,\ 2,\ 3, \\ \frac{d}{dt} \dot{x}_i &= \frac{\partial V}{\partial x_i}, \qquad i = 1,\ 2,\ 3, \end{aligned} \tag{3.26}$$

where x_i, $\dot{x}_i$ denote inertially fixed rectangular components of position and velocity, respectively. The rates of change dx_i/dt and $d\dot{x}_i/dt$ in (3.26) can be expressed as functions of the rates of change ds_k/dt of the six Keplerian elements, where s_k represents any of a, e, i, M, ω, or Ω. Thus

$$\sum_{k=1}^{6} \frac{\partial x_i}{\partial s_k} \cdot \frac{ds_k}{dt} = \frac{\partial x_i}{\partial s_k} \cdot \frac{ds_k}{dt} = \dot{x}_i, \qquad i = 1,\ 2,\ 3, \tag{3.27}$$

$$\sum_{k=1}^{6} \frac{\partial \dot{x}_i}{\partial s_k} \cdot \frac{ds_k}{dt} = \frac{\partial \dot{x}_i}{\partial s_k} \cdot \frac{ds_k}{dt} = \frac{\partial V}{\partial x_i}, \qquad i = 1,\ 2,\ 3, \tag{3.28}$$

where $\partial x_i/\partial s_k$ is obtained by differentiating (3.22) and (3.23) and $\partial \dot{x}_i/\partial s_k$ by differentiating (3.22) and (3.24). In the central formulas of (3.27) and (3.28) we have followed the convention that summation takes place over subscripts repeated in a product. The summation symbol will be omitted hereafter. In order to complete the conversion, for each element s_l in turn: (1) multiply (3.27) by $-\partial \dot{x}_i/\partial s_l$, (2) multiply (3.28) by $\partial x_i/\partial s_l$, and (3) add the resulting equations together. Thus

$$-\frac{\partial \dot{x}_i}{\partial s_l} \cdot \frac{\partial x_i}{\partial s_k} \cdot \frac{ds_k}{dt} + \frac{\partial x_i}{\partial s_l} \cdot \frac{\partial \dot{x}_i}{\partial s_k} \cdot \frac{ds_k}{dt} = -\frac{\partial \dot{x}_i}{\partial s_l} \dot{x}_i + \frac{\partial x_i}{\partial s_l} \cdot \frac{\partial V}{\partial x_i} \tag{3.29}$$

or

$$[s_l,\ s_k] \frac{ds_k}{dt} = \frac{\partial F}{\partial s_l} \tag{3.30}$$

if we sum over k. Here

$$[s_l, s_k] = \frac{\partial x_i}{\partial s_l} \cdot \frac{\partial \dot{x}_i}{\partial s_k} - \frac{\partial \dot{x}_i}{\partial s_l} \cdot \frac{\partial x_i}{\partial s_k}, \tag{3.31}$$

which is known as Lagrange's brackets, and

$$F = V - T. \tag{3.32}$$

F is known as the force function; it is the negative of the total energy as used in physics. V is the negative of the potential energy, and T is the kinetic energy. Thus, summing over i,

$$T = \tfrac{1}{2}\dot{x}_i \dot{x}_i. \tag{3.33}$$

The foregoing treatment was essentially first carried out by Lagrange. In celestial mechanics this treatment is customarily applied to the expression of the x_i, $\dot{x}_i$ in terms of the time t and the set of Keplerian elements at another time t_0 called the epoch. In this situation the kinetic energy T and the central term μ/r of V do not appear on the right. There are now two principal problems: (1) the formulation of the Lagrangian brackets $[s_l, s_k]$, and (2) the transformation of the potential V from rectangular or polar coordinates, such as (1.31), to Keplerian elements.

The form of (3.31) indicates that $[s_l, s_k]$ is the negative of $[s_k, s_l]$ and that $[s_k, s_k]$ vanishes, so there are fifteen different Lagrangian brackets to be determined by differentiating (3.22). A property of the Lagrangian brackets that facilitates their evaluation is their time invariance. Thus,

$$\begin{aligned}
\frac{\partial}{\partial t}[s_l, s_k] &= \frac{\partial^2 x_i}{\partial s_l \partial t} \cdot \frac{\partial \dot{x}_i}{\partial s_k} + \frac{\partial x_i}{\partial s_l} \cdot \frac{\partial^2 \dot{x}_i}{\partial s_k \partial t} - \frac{\partial^2 \dot{x}_i}{\partial s_l \partial t} \cdot \frac{\partial x_i}{\partial s_k} - \frac{\partial \dot{x}_i}{\partial s_l} \cdot \frac{\partial^2 x_i}{\partial s_k \partial t} \\
&= \frac{\partial}{\partial s_l}\left[\frac{\partial x_i}{\partial t} \cdot \frac{\partial \dot{x}_i}{\partial s_k} - \frac{\partial x_i}{\partial s_k} \cdot \frac{\partial \dot{x}_i}{\partial t}\right] - \frac{\partial}{\partial s_k}\left[\frac{\partial x_i}{\partial t} \cdot \frac{\partial \dot{x}_i}{\partial s_l} - \frac{\partial x_i}{\partial s_l} \cdot \frac{\partial \dot{x}_i}{\partial t}\right] \\
&= \frac{\partial}{\partial s_l}\left[\dot{x}_i \frac{\partial \dot{x}_i}{\partial s_k} - \frac{\partial x_i}{\partial s_k} \ddot{x}_i\right] - \frac{\partial}{\partial s_k}\left[\dot{x}_i \frac{\partial \dot{x}_i}{\partial s_l} - \frac{\partial x_i}{\partial s_l} \ddot{x}_i\right] \\
&= \frac{\partial}{\partial s_l}\left[\frac{1}{2}\frac{\partial(v^2)}{\partial s_k} - \frac{\partial x_i}{\partial s_k} \cdot \frac{\partial(\mu/r)}{\partial x_i}\right] - \frac{\partial}{\partial s_k}\left[\frac{1}{2}\frac{\partial(v^2)}{\partial s_l} - \frac{\partial x_i}{\partial s_l} \cdot \frac{\partial(\mu/r)}{\partial x_i}\right] \\
&= \frac{1}{2}\frac{\partial^2(v^2)}{\partial s_l \partial s_k} - \frac{\partial^2(\mu/r)}{\partial s_l \partial s_k} - \frac{1}{2}\frac{\partial^2(v^2)}{\partial s_l \partial s_k} + \frac{\partial^2(\mu/r)}{\partial s_l \partial s_k} = 0.
\end{aligned} \tag{3.34}$$

Hence the $\mathbf{q}$ and $\dot{\mathbf{q}}$ that appear in the expression (3.22) for $\mathbf{x}$ and $\dot{\mathbf{x}}$, and their derivatives, can be evaluated at a convenient point, such as perigee, where E is zero. Evaluating $\mathbf{q}$ and $\dot{\mathbf{q}}$ at perigee, we get from (3.22)–(3.24) in (3.31)

$$\text{(a)}\quad [s_l, s_k] = \left(\frac{\partial r_{i1}}{\partial s_l}\cdot\frac{\partial r_{i2}}{\partial s_k} - \frac{\partial r_{i2}}{\partial s_l}\cdot\frac{\partial r_{i1}}{\partial s_k}\right) na^2\sqrt{1-e^2}$$

$$\text{if } s_l = \Omega,\ i, \text{ or } \omega \quad \text{and} \quad \text{if } s_k = \Omega,\ i, \text{ or } \omega;$$

$$\text{(b)}\quad [s_l, s_k] = a(1-e)\frac{\partial r_{i1}}{\partial s_l}\left(r_{i1}\frac{\partial \dot{q}_1}{\partial s_k} + r_{i2}\frac{\partial \dot{q}_2}{\partial s_k}\right)$$

$$- \frac{\sqrt{1-e^2}\,na}{1-e}\cdot\frac{\partial r_{i2}}{\partial s_l}\left(r_{i1}\frac{\partial q_1}{\partial s_k} + r_{i2}\frac{\partial q_2}{\partial s_k}\right) \tag{3.35}$$

$$\text{if } s_l = \Omega,\ i, \text{ or } \omega \quad \text{and} \quad s_k = a,\ e,\ M;$$

$$\text{(c)}\quad [s_l, s_k] = r_{i1}r_{i1}\left(\frac{\partial q_1}{\partial s_l}\frac{\partial \dot{q}_1}{\partial s_k} - \frac{\partial q_1}{\partial s_k}\frac{\partial \dot{q}_1}{\partial s_l}\right) + r_{i1}r_{i2}\left(\frac{\partial q_1}{\partial s_l}\frac{\partial \dot{q}_2}{\partial s_k} - \frac{\partial q_1}{\partial s_k}\frac{\partial \dot{q}_2}{\partial s_l}\right)$$

$$+ r_{i2}r_{i1}\left(\frac{\partial q_2}{\partial s_l}\frac{\partial \dot{q}_1}{\partial s_k} - \frac{\partial q_2}{\partial s_k}\frac{\partial \dot{q}_1}{\partial s_l}\right) + r_{i2}r_{i2}\left(\frac{\partial q_2}{\partial s_l}\frac{\partial \dot{q}_2}{\partial s_k} - \frac{\partial q_2}{\partial s_k}\frac{\partial \dot{q}_2}{\partial s_l}\right)$$

$$\text{if } s_l = a,\ e, \text{ or } M \quad \text{and} \quad s_k = a,\ e, \text{ or } M.$$

In (3.35), r_{i1} and r_{i2} are elements of $\mathbf{R}_{xq}$; see (2.32). We have, for example, for $[\Omega, i]$ from (a) of (3.35) and (2.32),

$$\begin{aligned}
[\Omega, i] &= \left(\frac{\partial r_{i1}}{\partial \Omega}\cdot\frac{\partial r_{i2}}{\partial i} - \frac{\partial r_{i2}}{\partial \Omega}\cdot\frac{\partial r_{i1}}{\partial i}\right) na^2\sqrt{1-e^2} \\
&= [(-\sin\Omega\cos\omega - \cos\Omega\cos i\sin\omega)\sin\Omega\sin i\cos\omega \\
&\quad - (\cos\Omega\cos\omega - \sin\Omega\cos i\sin\omega)\cos\Omega\sin i\cos\omega \\
&\quad - (\sin\Omega\sin\omega - \cos\Omega\cos i\cos\omega)\sin\Omega\sin i\sin\omega \\
&\quad - (\cos\Omega\sin\omega + \sin\Omega\cos i\cos\omega)\cos\Omega\sin i\sin\omega] na^2\sqrt{1-e^2} \\
&= -na^2\sqrt{1-e^2}\sin i.
\end{aligned} \tag{3.36}$$

The complete set of nonzero results is

$$\begin{aligned}
[\Omega, i] &= -[i, \Omega] = -na^2(1-e^2)^{1/2}\sin i, \\
[\Omega, a] &= -[a, \Omega] = (1-e^2)^{1/2}\cos i\, na/2, \\
[\Omega, e] &= -[e, \Omega] = -na^2 e\cos i/(1-e^2)^{1/2}, \\
[\omega, a] &= -[a, \omega] = (1-e^2)^{1/2} na/2, \\
[\omega, e] &= -[e, \omega] = -na^2 e/(1-e^2)^{1/2}, \\
[a, M] &= -[M, a] = -na/2.
\end{aligned} \tag{3.37}$$

The substitution of expressions (3.37) in (3.30) and the solution of the six simultaneous equations for the ds_k/dt yield

$$
\begin{aligned}
\frac{da}{dt} &= \frac{2}{na}\frac{\partial F}{\partial M}, \\
\frac{de}{dt} &= \frac{1-e^2}{na^2e}\frac{\partial F}{\partial M} - \frac{(1-e^2)^{1/2}}{na^2e}\cdot\frac{\partial F}{\partial \omega}, \\
\frac{d\omega}{dt} &= -\frac{\cos i}{na^2(1-e^2)^{1/2}\sin i}\frac{\partial F}{\partial i} + \frac{(1-e^2)^{1/2}}{na^2e}\cdot\frac{\partial F}{\partial e}, \\
\frac{di}{dt} &= \frac{\cos i}{na^2(1-e^2)^{1/2}\sin i}\frac{\partial F}{\partial \omega} - \frac{1}{na^2(1-e^2)^{1/2}\sin i}\frac{\partial F}{\partial \Omega}, \\
\frac{d\Omega}{dt} &= \frac{1}{na^2(1-e^2)^{1/2}\sin i}\frac{\partial F}{\partial i}, \\
\frac{dM}{dt} &= -\frac{1-e^2}{na^2e}\frac{\partial F}{\partial e} - \frac{2}{na}\frac{\partial F}{\partial a}.
\end{aligned}
\tag{3.38}
$$

It is customary to express the force function as

$$
\begin{aligned}
F &= \frac{\mu}{r} + R - T \\
&= \frac{\mu}{2a} + R,
\end{aligned}
\tag{3.39}
$$

from (3.25). The function R, comprising all terms of V except the central term, is known as the disturbing function. Hence in all equations of (3.38) F can be replaced by R except in the last, which becomes, using (3.20),

$$
\frac{dM}{dt} = n - \frac{1-e^2}{na^2e}\cdot\frac{\partial R}{\partial e} - \frac{2}{na}\cdot\frac{\partial R}{\partial a}. \tag{3.40}
$$

The symmetries and similarities of the brackets in (3.37) suggest that further simplifications may be made by change of variables from Keplerian. Let us try to find a set $L,\ G,\ H$ such that

$$
\begin{aligned}
&[M,\ L] = 1, \qquad &&[M,\ G] = 0, \qquad &&[M,\ H] = 0, \\
&[\omega,\ L] = 0, &&[\omega,\ G] = 1, &&[\omega,\ H] = 0, \\
&[\Omega,\ L] = 0, &&[\Omega,\ G] = 0, &&[\Omega,\ H] = 1.
\end{aligned}
\tag{3.41}
$$

The only nonzero bracket in (3.37) involving the inclination i is the first. From (3.31) we must have

$$[\Omega, H]\frac{\partial H}{\partial i} = [\Omega, i] = -na^2(1 - e^2)^{1/2} \sin i, \tag{3.42}$$

whence

$$H = na^2(1 - e^2)^{1/2} \cos i. \tag{3.43}$$

As a check, we find (remembering n is $\mu^{1/2}a^{-3/2}$)

$$[\Omega, H]\frac{\partial H}{\partial e} = [\Omega, e], \qquad [\Omega, H]\frac{\partial H}{\partial a} = [\Omega, a]. \tag{3.44}$$

Similarly from $[\omega, G]\, \partial G/\partial e$ and $[M, L]\, \partial L/\partial a$, we have

$$G = na^2(1 - e^2)^{1/2} \tag{3.45}$$

and

$$L = na^2 = \mu^{1/2}a^{1/2}. \tag{3.46}$$

We thus obtain (3.47), the somewhat simpler Delaunay equations,

$$\begin{aligned} \frac{dL}{dt} &= \frac{\partial F}{\partial M}, & \frac{dM}{dt} &= -\frac{\partial F}{\partial L}, \\ \frac{dG}{dt} &= \frac{\partial F}{\partial \omega}, & \frac{d\omega}{dt} &= -\frac{\partial F}{\partial G}, \\ \frac{dH}{dt} &= \frac{\partial F}{\partial \Omega}, & \frac{d\Omega}{dt} &= -\frac{\partial F}{\partial H}. \end{aligned} \tag{3.47}$$

In using Delaunay variables the notation M, ω, Ω is usually replaced by the notation l, g, h [not to be confused with the h defined by (3.4) and (3.12)].

3.3. Conversion of Spherical Harmonic Disturbing Function

In order to convert the spherical harmonic potential (1.31) to Keplerian elements, we require some trigonometric identities, such as

$$\begin{aligned} \cos mx &= \operatorname{Re} \exp(mjx) = \operatorname{Re}(\cos x + j \sin x)^m \\ &= \operatorname{Re} \sum_{s=0}^{m} \binom{m}{s} j^s \cos^{m-s} x \sin^s x, \end{aligned} \tag{3.48}$$

where Re denotes the real part, j is $\sqrt{-1}$, and $\binom{m}{s}$ is the binomial coefficient:

$$\binom{m}{s} = \frac{m!}{s!\,(m-s)!}, \tag{3.49}$$

$$\begin{aligned}\sin mx &= \text{Re}\,[-j \exp(mjx)] = \text{Re}\,[-j(\cos x + j \sin x)^m] \\ &= \text{Re} \sum_{s=0}^{m} \binom{m}{s} j^{s-1} \cos^{m-s} x \sin^s x,\end{aligned} \tag{3.50}$$

$$\begin{aligned}\sin^a x \cos^b x &= \left[-\frac{j}{2}(e^{jx} - e^{-jx})\right]^a \left[\frac{1}{2}(e^{jx} + e^{-jx})\right]^b \\ &= \frac{(-1)^a j^a}{2^a} \sum_{c=0}^{a} \binom{a}{c} e^{(a-c)jx}(-1)^c e^{-cjx} \cdot \frac{1}{2^b} \sum_{d=0}^{b} \binom{b}{d} \cdot e^{(b-d)jx} e^{-djx} \\ &= \frac{(-1)^a j^a}{2^{a+b}} \sum_{c=0}^{a} \sum_{d=0}^{b} \binom{a}{c} \binom{b}{d} e^{(a+b-2c-2d)jx}(-1)^c \\ &= \frac{(-1)^a j^a}{2^{a+b}} \sum_{c=0}^{a} \sum_{d=0}^{b} \binom{a}{c} \binom{b}{d} (-1)^c \\ &\quad \times [\cos(a+b-2c-2d)x \\ &\quad + j \sin(a+b-2c-2d)x],\end{aligned} \tag{3.51}$$

$$\begin{aligned}\cos a \cos b &= \tfrac{1}{2} \cos(a+b) + \tfrac{1}{2} \cos(a-b), \\ \sin a \sin b &= -\tfrac{1}{2} \cos(a+b) + \tfrac{1}{2} \cos(a-b), \\ \sin a \cos b &= \tfrac{1}{2} \sin(a+b) + \tfrac{1}{2} \sin(a-b), \\ \cos a \sin b &= \tfrac{1}{2} \sin(a+b) - \tfrac{1}{2} \sin(a-b).\end{aligned} \tag{3.52}$$

Let a particular term of (1.31) be

$$V_{lm} = \frac{\mu a_e^l}{r^{l+1}} P_{lm}(\sin \phi)(C_{lm} \cos m\lambda + S_{lm} \sin m\lambda). \tag{3.53}$$

We have made the C_{lm}, S_{lm} nondimensional by applying the factor μa_e^l, where a_e is the equatorial radius of the earth. We then substitute $[m(\alpha - \Omega) + m(\Omega - \theta)]$ for $m\lambda$, where α is right ascension and θ is Greenwich Sidereal Time:

$$\begin{aligned}\cos m\lambda &= \cos m(\alpha - \Omega) \cos m(\Omega - \theta) - \sin m(\alpha - \Omega) \sin m(\Omega - \theta), \\ \sin m\lambda &= \sin m(\alpha - \Omega) \cos m(\Omega - \theta) + \cos m(\alpha - \Omega) \sin m(\Omega - \theta).\end{aligned} \tag{3.54}$$

In the spherical triangle formed by the orbit, the equator, and the satellite meridian (see Figure 4) we have

$$\cos(\omega + f) = \cos(\alpha - \Omega)\cos\phi + \sin(\alpha - \Omega)\sin\phi\cos\pi/2,$$
$$\cos\phi = \cos(\omega + f)\cos(\alpha - \Omega) + \sin(\omega + f)\sin(\alpha - \Omega)\cos i,$$

whence

$$\cos(\alpha - \Omega) = \cos(\omega + f)/\cos\phi, \tag{3.55}$$
$$\sin(\alpha - \Omega) = \sin(\omega + f)\cos i/\cos\phi,$$

and

$$\sin\phi = \sin i\sin(\omega + f). \tag{3.56}$$

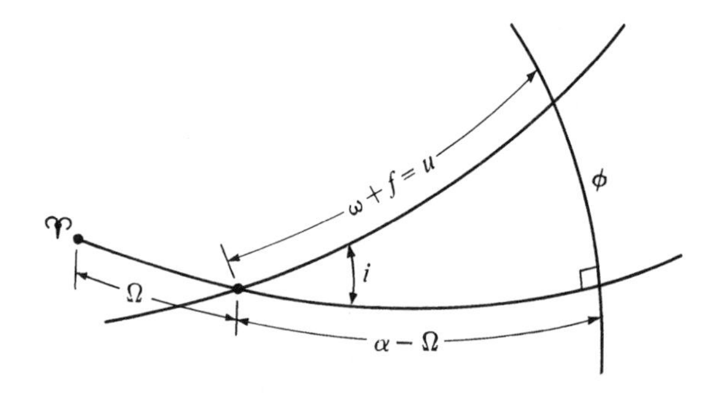

FIGURE 4. *Orbit-equator-meridian triangle.*

If we apply (3.48) and (3.50) to the $(\alpha - \Omega)$ functions in (3.54) and substitute (3.55) therein, we get

$$\begin{aligned}\cos m\lambda &= \operatorname{Re}\sum_{s=0}^{m}\binom{m}{s}j^{s}\frac{\cos^{m-s}(\omega + f)\sin^{s}(\omega + f)\cos^{s} i}{\cos^{m}\phi}\\ &\qquad\times[\cos m(\Omega - \theta) + j\sin m(\Omega - \theta)],\\ \sin m\lambda &= \operatorname{Re}\sum_{s=0}^{m}\binom{m}{s}j^{s}\frac{\cos^{m-s}(\omega + f)\sin^{s}(\omega + f)\cos^{s} i}{\cos^{m}\phi}\\ &\qquad\times[\sin m(\Omega - \theta) - j\cos m(\Omega - \theta)].\end{aligned} \tag{3.57}$$

If we substitute (3.56) for $\sin\phi$ in (1.30), which is the definition of P_{lm}, and then substitute both (1.30) and (3.57) in (3.53), by cancelling out the

$\cos^m \phi$'s, we have

$$V_{lm} = \frac{\mu a_e^l}{r^{l+1}} \sum_{t=0}^{k} T_{lmt} \sin^{l-m-2t} i \ \mathrm{Re}\,[(C_{lm} - jS_{lm}) \cos m(\Omega - \theta) + (S_{lm} + jC_{lm}) \sin m(\Omega - \theta)] \sum_{s=0}^{m} \binom{m}{s} j^s \sin^{l-m-2t+s} (\omega + f) \times \cos^{m-s} (\omega + f) \cos^s i, \quad (3.58)$$

where k is the integer part of $(l - m)/2$.

On applying (3.51) to (3.58), with $a = l - m - 2t + s$, and $b = m - s$,

$$V_{lm} = \frac{\mu a_e^l}{r^{l+1}} \sum_{t=0}^{k} T_{lmt} \sin^{l-m-2t} i \ \mathrm{Re}\,[(C_{lm} - jS_{lm}) \cos m(\Omega - \theta) + (S_{lm} + jC_{lm}) \sin m(\Omega - \theta)] \sum_{s=0}^{m} \binom{m}{s} j^s \cos^s i \frac{[-j]^{l-m-2t+s}}{2^{l-2t}}$$
$$\times \sum_{c=0}^{l-m-2t+s} \sum_{d=0}^{m-s} \binom{l-m-2t+s}{c} \binom{m-s}{d} (-1)^c$$
$$\times [\cos (l - 2t - 2c - 2d)(\omega + f) + j \sin (l - 2t - 2c - 2d)(\omega + f)]. \quad (3.59)$$

By applying (3.52) to the products of $(\Omega - \theta)$ and $(\omega + f)$ trigonometric functions in (3.59), and dropping any term with an odd power of j as a coefficient (since V_{lm} is real, such a term has another term cancelling it out), we have

$$V_{lm} = \frac{\mu a_e^l}{r^{l+1}} \sum_{t=0}^{k} T_{lmt} \sin^{l-m-2t} i (-1)^{k+t} \sum_{s=0}^{m} \binom{m}{s} \frac{\cos^s i}{2^{l-2t}}$$
$$\times \sum_{c=0}^{l-m-2t+s} \sum_{d=0}^{m-s} \binom{l-m-2t+s}{c} \binom{m-s}{d} (-1)^c$$
$$\times \left\{ \begin{bmatrix} C_{lm} \\ -S_{lm} \end{bmatrix}_{l-m \text{ odd}}^{l-m \text{ even}} \cos [(l - 2t - 2c - 2d)(\omega + f) + m(\Omega - \theta)] + \begin{bmatrix} S_{lm} \\ C_{lm} \end{bmatrix}_{l-m \text{ odd}}^{l-m \text{ even}} \sin [(l - 2t - 2c - 2d)(\omega + f) + m(\Omega - \theta)] \right\}. \quad (3.60)$$

It is desirable to transform (3.60) so that terms of the same argument $[(l - 2p)(\omega + f) + m(\Omega - \theta)]$ are collected together. By substituting p for $(t + c + d)$ necessitates, in turn, the elimination of one subscript from the factors. Putting $p - t - c$ in place of d seems most convenient. The limits of the d summation place limits on the possible values of c, which turn out to be simply those making the binomial coefficients nonzero. In addition, $t \leq p$.

The expression for V_{lm} thus becomes

$$V_{lm} = \frac{\mu a_e^l}{r^{l+1}} \sum_{p=0}^{l} F_{lmp}(i) \left\{ \begin{bmatrix} C_{lm} \\ -S_{lm} \end{bmatrix}_{l-m \text{ odd}}^{l-m \text{ even}} \cos\,[(l-2p)(\omega + f) + m(\Omega - \theta)] + \begin{bmatrix} S_{lm} \\ C_{lm} \end{bmatrix}_{l-m \text{ odd}}^{l-m \text{ even}} \sin\,[(l-2p)(\omega + f) + m(\Omega - \theta)] \right\}, \tag{3.61}$$

where, substituting from (1.29),

$$F_{lmp}(i) = \sum_t \frac{(2l-2t)!}{t!\,(l-t)!\,(l-m-2t)!\,2^{2l-2t}} \sin^{l-m-2t} i \times \sum_{s=0}^{m} \binom{m}{s} \cos^s i \sum_c \binom{l-m-2t+s}{c} \binom{m-s}{p-t-c} (-1)^{c-k}. \tag{3.62}$$

Here k is the integer part of $(l - m)/2$, t is summed from 0 to the lesser of p or k, and c is summed over all values making the binomial coefficients nonzero. A formula such as (3.62) is convenient only for computer use. For hand calculations, a table is better; expressions for $F_{lmp}(i)$ up to $lmp = 444$ are given in Table 1.

TABLE 1

Inclination Functions $F_{lmp}(i)$ from Equation (3.62)

l	m	p	$F_{lmp}(i)$
2	0	0	$-3 \sin^2 i/8$
2	0	1	$3 \sin^2 i/4 - 1/2$
2	0	2	$-3 \sin^2 i/8$
2	1	0	$3 \sin i(1 + \cos i)/4$
2	1	1	$-3 \sin i \cos i/2$
2	1	2	$-3 \sin i(1 - \cos i)/4$
2	2	0	$3(1 + \cos i)^2/4$
2	2	1	$3 \sin^2 i/2$
2	2	2	$3(1 - \cos i)^2/4$
3	0	0	$-5 \sin^3 i/16$
3	0	1	$15 \sin^3 i/16 - 3 \sin i/4$
3	0	2	$-15 \sin^3 i/16 + 3 \sin i/4$
3	0	3	$5 \sin^3 i/16$
3	1	0	$-15 \sin^2 i(1 + \cos i)/16$

TABLE 1 (*Continued*)

l	m	p	$F_{lmp}(i)$
3	1	1	$15 \sin^2 i(1 + 3 \cos i)/16 - 3(1 + \cos i)/4$
3	1	2	$15 \sin^2 i(1 - 3 \cos i)/16 - 3(1 - \cos i)/4$
3	1	3	$-15 \sin^2 i(1 - \cos i)/16$
3	2	0	$15 \sin i(1 + \cos i)^2/8$
3	2	1	$15 \sin i(1 - 2 \cos i - 3 \cos^2 i)/8$
3	2	2	$-15 \sin i (1 + 2 \cos i - 3 \cos^2 i)/8$
3	2	3	$-15 \sin i(1 - \cos i)^2/8$
3	3	0	$15(1 + \cos i)^3/8$
3	3	1	$45 \sin^2 i(1 + \cos i)/8$
3	3	2	$45 \sin^2 i(1 - \cos i)/8$
3	3	3	$15(1 - \cos i)^3/8$
4	0	0	$35 \sin^4 i/128$
4	0	1	$-35 \sin^4 i/32 + 15 \sin^2 i/16$
4	0	2	$(105/64) \sin^4 i - (15/8) \sin^2 i + 3/8$
4	0	3	$-(35/32) \sin^4 i + (15/16) \sin^2 i$
4	0	4	$(35/128) \sin^4 i$
4	1	0	$-(35/32) \sin^3 i(1 + \cos i)$
4	1	1	$(35/16) \sin^3 i(1 + 2 \cos i) - (15/8)(1 + \cos i) \sin i$
4	1	2	$\cos i(15 \sin i/4 - 105 \sin^3 i/16)$
4	1	3	$-(35/16) \sin^3 i(1 - 2 \cos i) + (15/8) \sin i(1 - \cos i)$
4	1	4	$(35/32) \sin^3 i(1 - \cos i)$
4	2	0	$-(105/32) \sin i (1 + \cos i)^2$
4	2	1	$(105/8) \sin^2 i \cos i(1 + \cos i) - (15/8)(1 + \cos i)^2$
4	2	2	$(105/16) \sin^2 i(1 - 3 \cos^2 i) + (15/4) \sin^2 i$
4	2	3	$-(105/8) \sin^2 i \cos i(1 - \cos i) - (15/8)(1 - \cos i)^2$
4	2	4	$-(105/32) \sin^2 i(1 - \cos i)^2$
4	3	0	$(105/16) \sin i(1 + \cos i)^3$
4	3	1	$(105/8) \sin i (1 - 3 \cos^2 i - 2 \cos^3 i)$
4	3	2	$-(315/8) \sin^3 i \cos i$
4	3	3	$-(105/8) \sin i(1 - 3 \cos^2 i + 2 \cos^3 i)$
4	3	4	$-(105/16) \sin i(1 - \cos i)^3$
4	4	0	$(105/16)(1 + \cos i)^4$
4	4	1	$(105/4) \sin^2 i(1 + \cos i)^2$
4	4	2	$(315/8) \sin^4 i$
4	4	3	$(105/4) \sin^2 i(1 - \cos i)^2$
4	4	4	$(105/16)(1 - \cos i)^4$

The final transformation necessary to obtain a disturbing function consistent with (3.38) is to replace r and f in (3.61) by a, M, and e, that is, to make the replacement

$$\frac{1}{r^{l+1}}\begin{bmatrix}\cos\\ \sin\end{bmatrix}[(l - 2p)(\omega + f) + m(\Omega - \theta)]$$

$$= \frac{1}{a^{l+1}} \sum_q G_{lpq}(e)\begin{bmatrix}\cos\\ \sin\end{bmatrix}[(l - 2p)\omega + (l - 2p + q)M + m(\Omega - \theta)].$$

In order to obtain long period terms—those terms from which M is absent—we can average (3.61) with respect to M, that is, we can integrate with respect to M from 0 to 2π and then divide by 2π. We then have from (3.4), (3.12), and (3.20),

$$dM = \frac{r^2\, df}{a^2(1-e^2)^{1/2}}, \tag{3.63}$$

and from (3.9),

$$r^2/r^{l+1} = [(1 + e\cos f)/a(1-e^2)]^{l-1}. \tag{3.64}$$

Binomially expanding $(1 + e\cos f)^{l-1}$, applying (3.51) and (3.52), and omitting imaginary terms, we get

$$\begin{aligned}
\frac{1}{2\pi}\int_0^{2\pi} \frac{1}{r^{l+1}} \begin{bmatrix}\cos\\ \sin\end{bmatrix} [(l-2p)(\omega+f) + m(\Omega-\theta)]\, dM \\
= \frac{1}{2\pi}\int_0^{2\pi} \frac{1}{a^{l+1}(1-e^2)^{l-(1/2)}} \sum_{b=0}^{l-1} \binom{l-1}{b}\left(\frac{e}{2}\right)^b \\
\times \sum_{d=0}^{b}\binom{b}{d}\frac{1}{2}\left\{\begin{bmatrix}\cos\\ \sin\end{bmatrix}[(l-2p)\omega + (l+b-2p-2d)f + m(\Omega-\theta)]\right. \\
\left. + \begin{bmatrix}\cos\\ \sin\end{bmatrix}[(l-2p)\omega + (l-b-2p+2d)f + m(\Omega-\theta)]\right\} df \\
= \frac{1}{a^{l+1}} G_{lp(2p-l)}(e) \begin{bmatrix}\cos\\ \sin\end{bmatrix}[(l-2p)\omega + m(\Omega-\theta)],
\end{aligned} \tag{3.65}$$

where

$$G_{lp(2p-l)}(e) = \frac{1}{(1-e^2)^{l-(1/2)}} \sum_{d=0}^{p'-1} \binom{l-1}{2d+l-2p'}\binom{2d+l-2p'}{d}\left(\frac{e}{2}\right)^{2d+l-2p'}, \tag{3.66}$$

in which

$$p' = p \quad \text{for} \quad p \le l/2,$$
$$p' = l - p \quad \text{for} \quad p \ge l/2.$$

The "$\frac{1}{2}$" factor inside the d summation of (3.65) has disappeared because the two terms that satisfy the condition for long period variation $(l-2p) \pm (b-2d) = 0$, are symmetrically placed coefficients in the binomial expansion, and hence can be combined by making the one substitution $(2d + l - 2p')$ for b in (3.66).

For the short period terms, $l - 2p + q \neq 0$, the development of $G_{lpq}(e)$ is much more complicated; we merely quote the result of one solution (Tisserand, 1889, p. 256):

$$G_{lpq}(e) = (-1)^{|q|}(1 + \beta^2)^l \beta^{|q|} \sum_{k=0}^{\infty} P_{lpqk} Q_{lpqk} \beta^{2k}, \tag{3.67}$$

where

$$\beta = \frac{e}{1 + \sqrt{1 - e^2}};$$

$$P_{lpqk} = \sum_{r=0}^{h} \binom{2p' - 2l}{h - r} \frac{(-1)^r}{r!} \left(\frac{(l - 2p' + q')e}{2\beta}\right)^r,$$
$$h = k + q', \quad q' > 0; \quad h = k, \quad q' < 0; \tag{3.68}$$

and

$$Q_{lpqk} = \sum_{r=0}^{h} \binom{-2p'}{h - r} \frac{1}{r!} \left(\frac{(l - 2p' + q')e}{2\beta}\right)^r,$$
$$h = k, \quad q' > 0; \quad h = k - q', \quad q' < 0; \tag{3.69}$$
$$p' = p, \quad q' = q \text{ for } p \leq l/2; \quad p' = l - p, \quad q' = -q \text{ for } p > l/2.$$

Expressions for $G_{lpq}(e)$ up to $lpq = 442$ are given in Table 2, which is based on the more extensive tables of Cayley (1861).

The final result for the transformation of V_{lm} in spherical coordinates, (3.53), to orbital coordinates can thus be expressed by

$$V_{lm} = \frac{\mu a_e^l}{a^{l+1}} \sum_{p=0}^{l} F_{lmp}(i) \sum_{q=-\infty}^{\infty} G_{lpq}(e) S_{lmpq}(\omega, M, \Omega, \theta), \tag{3.70}$$

where

$$S_{lmpq} = \begin{bmatrix} C_{lm} \\ -S_{lm} \end{bmatrix}_{l-m \text{ odd}}^{l-m \text{ even}} \cos\,[(l - 2p)\omega + (l - 2p + q)M + m(\Omega - \theta)]$$
$$+ \begin{bmatrix} S_{lm} \\ C_{lm} \end{bmatrix}_{l-m \text{ odd}}^{l-m \text{ even}} \sin\,[(l - 2p)\omega + (l - 2p + q)M + m(\Omega - \theta)]. \tag{3.71}$$

3.4. Linear Perturbations

As discussed in Chapter 1, the term of the gravitational field which will dominate the disturbing function R is V_{20}, since C_{20} (or $-J_2$) is at least

TABLE 2

Eccentricity Functions $G_{lpq}(e)$ from Equations (3.66)–(3.69) *or the Tables of Cayley* (1861)

l	p	q	l	p	q	$G_{lpq}(e)$
2	0	−2	2	2	2	0
2	0	−1	2	2	1	$-e/2 + e^3/16 + \cdots$
2	0	0	2	2	0	$1 - 5e^2/2 + 13e^4/16 + \cdots$
2	0	1	2	2	−1	$7e/2 - 123e^3/16 + \cdots$
2	0	2	2	2	−2	$17e^2/2 - 115e^4/6 + \cdots$
2	1	−2	2	1	2	$9e^2/4 + 7e^4/4 + \cdots$
2	1	−1	2	1	1	$3e/2 + 27e^3/16 + \cdots$
			2	1	0	$(1 - e^2)^{-3/2}$
3	0	−2	3	3	2	$e^2/8 + e^4/48 + \cdots$
3	0	−1	3	3	1	$-e + 5e^3/4 + \cdots$
3	0	0	3	3	0	$1 - 6e^2 + 423e^4/64 + \cdots$
3	0	1	3	3	−1	$5e - 22e^3 + \cdots$
3	0	2	3	3	−2	$127e^2/8 - 3065e^4/48 + \cdots$
3	1	−2	3	2	2	$11e^2/8 + 49e^4/16 + \cdots$
3	1	−1	3	2	1	$e(1 - e^2)^{-5/2}$
3	1	0	3	2	0	$1 + 2e^2 + 239e^4/64 + \cdots$
3	1	1	3	2	−1	$3e + 11e^3/4 + \cdots$
3	1	2	3	2	−2	$53e^2/8 + 39e^4/16 + \cdots$
4	0	−2	4	4	2	$e^2/2 - e^4/3 + \cdots$
4	0	−1	4	4	1	$-3e/2 + 75e^3/16 + \cdots$
4	0	0	4	4	0	$1 - 11e^2 + 199e^4/8 + \cdots$
4	0	1	4	4	−1	$13e/2 - 765e^3/16 + \cdots$
4	0	2	4	4	−2	$51e^2/2 - 321e^4/2 + \cdots$
4	1	−2	4	3	2	$(3e^2/4)(1 - e^2)^{-7/2}$
4	1	−1	4	3	1	$e/2 + 33e^2/16 + \cdots$
4	1	0	4	3	0	$1 + e^2 + 65e^4/16 + \cdots$
4	1	1	4	3	−1	$9e/2 - 3e^3/16 + \cdots$
4	1	2	4	3	−2	$53e^2/4 - 179e^4/24 + \cdots$
4	2	−2	4	2	2	$5e^2 + 155e^4/12 + \cdots$
4	2	−1	4	2	1	$5e/2 + 135e^3/16 + \cdots$
			4	2	0	$(1 + 3e^2/2)(1 - e^2)^{-7/2}$

100 times as great as any other C_{lm}. Thus

$$V_{20} = \frac{\mu C_{20}}{a}\left(\frac{a_e}{a}\right)^2 \sum_{p}\sum_{q} F_{20p}(i)G_{2pq}(e) \cos\left[(2-2p)\omega + (2-2p+q)M\right]. \quad (3.72)$$

Assuming that the coefficients in V_{20} are of about the same magnitude, we should expect the terms in (3.72) that do not contain M—$(2 - 2p + q)$ is zero—to be of appreciably greater effect after integration than those which do contain M, since the latter will go through a full cycle in the relatively brief time it takes the satellite to complete a revolution. The summation limit in (3.66) indicates that terms with subscripts (p, q) of $(0, -2)$ and $(2, 2)$ do not exist, so the only term of V_{20} from which M is absent is

$$V_{2010} = \frac{\mu C_{20}}{a}\left(\frac{a_e}{a}\right)^2 F_{201}(i)G_{210}(e). \quad (3.73)$$

Evaluating F_{201} by (3.62) and G_{210} by (3.66), and using $\frac{\mu}{2a} + V_{2010}$ for F in the Lagrangian equations (3.38), we get

$$\begin{aligned}
\frac{da}{dt} &= 0,\\
\frac{de}{dt} &= 0,\\
\frac{d\omega}{dt} &= \frac{\mu C_{20}a_e^2}{n(1-e^2)^{1/2}a^5}\left[-\cot i\,\frac{\partial F_{201}}{\partial i}G_{210} + \frac{(1-e^2)}{e}F_{201}\frac{\partial G_{210}}{\partial e}\right]\\
&= \frac{3nC_{20}a_e^2}{4(1-e^2)^2a^2}\left[1 - 5\cos^2 i\right],\\
\frac{di}{dt} &= 0, \qquad\qquad (3.74)\\
\frac{d\Omega}{dt} &= \frac{\mu C_{20}a_e^2G_{210}}{n(1-e^2)^{1/2}a^5 \sin i}\cdot\frac{\partial F_{201}}{\partial i}\\
&= \frac{3nC_{20}a_e^2}{2(1-e^2)^2a^2}\cos i,\\
\frac{dM}{dt} &= n + \frac{\mu C_{20}a_e^2F_{201}}{na^5}\left[-\frac{1-e^2}{e}\cdot\frac{\partial G_{210}}{\partial e} + 6G_{210}\right]\\
&= n - \frac{3nC_{20}a_e^2}{4(1-e^2)^{3/2}a^2}(3\cos^2 i - 1).
\end{aligned}$$

The value of C_{20} is about -0.0010827; on using typical geodetic satellite orbit specifications such as $e = 0.01$ and $a = 1.12a_e$, the above formulas yield

$$\begin{aligned}
\frac{d\omega}{dt} &\approx +3.55(5\cos^2 i - 1) \text{ degrees/day}, \\
\frac{d\Omega}{dt} &\approx -6.70 \cos i \text{ degrees/day}, \\
\frac{dM}{dt} &\approx 14.37 + 0.0093(3\cos^2 i - 1) \text{ revolutions/day}.
\end{aligned} \tag{3.75}$$

It is an observed fact that the secular motions such as in (3.75) are the dominant perturbation of geodetically useful satellites—that is, those high enough not to suffer excessive drag, but low enough to be perceptibly perturbed by the variations of the earth's gravitational field. Hence the first approximation to an integration of the equations of motion (3.38) for the effect of a particular disturbing function R, such as a potential field term V_{lm}, will be one that assumes that the only variations with time of the elements on the right side of the equations are the secular rates $\dot{\omega}$, $\dot{\Omega}$, $\dot{M}$, plus any other rates of change from outside the orbit, such as the rotation rate of the earth $\dot{\theta}$. Under this assumption, the integration of (3.38) with one term of (3.70), V_{lmpq}, as the disturbing function will be

$$\begin{aligned}
\Delta a_{lmpq} &= \mu a_e^l \frac{2F_{lmp}G_{lpq}(l-2p+q)S_{lmpq}}{na^{l+2}[(l-2p)\dot{\omega} + (l-2p+q)\dot{M} + m(\dot{\Omega} - \dot{\theta})]}, \\
\Delta e_{lmpq} &= \mu a_e^l \frac{F_{lmp}G_{lpq}(1-e^2)^{1/2}[(1-e^2)^{1/2}(l-2p+q) - (l-2p)]S_{lmpq}}{na^{l+3}e[(l-2p)\dot{\omega} + (l-2p+q)\dot{M} + m(\dot{\Omega} - \dot{\theta})]}, \\
\Delta \omega_{lmpq} &= \mu a_e^l \frac{[(1-e^2)^{1/2}e^{-1}F_{lmp}(\partial G_{lpq}/\partial e) - \cot i(1-e^2)^{-1/2}(\partial F_{lmp}/\partial i)G_{lpq}]\bar{S}_{lmpq}}{na^{l+3}[(l-2p)\dot{\omega} + (l-2p+q)\dot{M} + m(\dot{\Omega} - \dot{\theta})]}, \\
\Delta i_{lmpq} &= \mu a_e^l \frac{F_{lmp}G_{lpq}[(l-2p)\cos i - m]S_{lmpq}}{na^{l+3}(1-e^2)^{1/2}\sin i[(l-2p)\dot{\omega} + (l-2p+q)\dot{M} + m(\dot{\Omega} - \dot{\theta})]}, \\
\Delta \Omega_{lmpq} &= \mu a_e^l \frac{(\partial F_{lmp}/\partial i)G_{lpq}\bar{S}_{lmpq}}{na^{l+3}(1-e^2)^{1/2}\sin i[(l-2p)\dot{\omega} + (l-2p+q)\dot{M} + m(\dot{\Omega} - \dot{\theta})]}, \\
\Delta M_{lmpq} &= \mu a_e^l \frac{[-(1-e^2)e^{-1}(\partial G_{lpq}/\partial e) + 2(l+1)G_{lpq}]F_{lmp}\bar{S}_{lmpq}}{na^{l+3}[(l-2p)\dot{\omega} + (l-2p+q)\dot{M} + m(\dot{\Omega} - \dot{\theta})]},
\end{aligned} \tag{3.76}$$

where $\bar{S}_{lmpq}$ is the integral of S_{lmpq} with respect to its argument.

Specific examples of the perturbations (3.76), using (3.62), (3.66) and (3.71), are

1. The long period perturbation of the eccentricity by C_{30}, the "pear-shaped" term is

$$\Delta e_{301(-1)} + \Delta e_{3021} = -\frac{\mu a_e^3(1-e^2)^{1/2}}{na^6 e\dot{\omega}} \sum_{p=1}^{2} F_{30p} G_{3p(2p-3)} S_{30p(2p-3)}$$
$$= \frac{3\mu a_e^3 C_{30}(1-\frac{5}{4}\sin^2 i)}{2na^6(1-e^2)^2\dot{\omega}} \sin\omega; \qquad (3.77)$$

2. The semidaily perturbation of the mean anomaly by C_{22} and S_{22}, the "equatorial ellipticity," is

$$\Delta M_{2210} = \frac{[-(1-e^2)e^{-1}(\partial G_{210}/\partial e) + 6G_{210}]\mu a_e^2 F_{221} \bar{S}_{2210}}{na^5 2(\dot{\Omega} - \dot{\theta})}$$
$$= \frac{9\mu a_e^2 \sin^2 i}{4na^5(1-e^2)^{3/2}(\dot{\Omega}-\dot{\theta})}[C_{22}\sin 2(\Omega-\theta) - S_{22}\cos 2(\Omega-\theta)]. \qquad (3.78)$$

3.5. Nonlinear Perturbations

Since C_{20} is about 1000 times as big as the other gravitational field coefficients C_{lm}, S_{lm}, we should expect that a solution of the problem of a close satellite motion in which the effects of the C_{lm}, S_{lm}'s are described as linear perturbations in accordance with (3.76) would require that the effects of C_{20} should be described as nonlinear perturbations to order $(C_{20})^2$. That is, linear perturbations of the elements on the right of the equations of motion (3.38) due to C_{20} should be taken into account. If the problem is solved by numerical integration of (3.38) with a suitable integration interval—or, for that matter, by numerical integration of the rectilinear equations (3.26), —then such higher-order effects will be automatically taken care of. For example, writing the equations (3.38) as

$$\dot{s}_i = \dot{s}_i[s_j, t] \equiv \dot{s}_i\{a, e, i, M, \omega, \Omega, t\},$$

where s_i is any one of the Keplerian elements, the integration for a time-step Δt can be done by the standard fourth-order Runge-Kutta technique:

$$\begin{aligned} w_i &= \dot{s}_i[s_j(t), t]\,\Delta t, \\ x_i &= \dot{s}_i[s_j(t) + w_j/2, t + \Delta t/2]\,\Delta t, \\ y_i &= \dot{s}_i[s_j(t) + x_j'2, t + \Delta t/2]\,\Delta t, \\ z_i &= \dot{s}_i[s_j(t) + y_j/t + \Delta t]\,\Delta t, \\ s_i(t+\Delta t) &= s_i(t) + w_i/6 + x_i/3 + y_i/3 + z_i/6. \end{aligned} \qquad (3.79)$$

An excellent summary of numerical integration techniques applied to satellite orbits is given by Conte (1962). A solution more in keeping with the development thus far, however, would be one analytically developing (3.38). Therefore,

$$\begin{aligned}\dot{s}_i &= \dot{s}_i(\mathbf{s}_0) + \frac{\partial \dot{s}_i}{\partial s_j}\Delta_1 s_j \\ &= \frac{d}{dt}(\Delta_1 s_i) + \frac{d}{dt}(\Delta_2 s_i),\end{aligned}$$

where s_i is any of the Keplerian elements.

Subtracting out the first approximation, $d(\Delta_1 s_i)/dt$, leaves

$$\frac{d}{dt}(\Delta_2 s_i) = \frac{\partial \dot{s}_i}{\partial s_j}\Delta_1 s_j, \tag{3.80}$$

where $\Delta_1 s_j$ is found by (3.76). However, it would require considerable tedious algebra to apply (3.80) directly in this manner. It would be laborious even to get the one most important second-order perturbation—the $(C_{20})^2$ contribution to the secular motion—because the interactions of all periodic terms with themselves must be taken into account. Thus for the secular part of $\Delta_2 s_i$, we have

$$\Delta_2 \dot{s}_i = \left(\frac{\partial \dot{s}_{i20pq}}{\partial s_j}\Delta_1 s_{j20pq} - \text{periodic terms}\right), \tag{3.81}$$

where summation is over j, p, and q. In order to apply (3.81) to the evaluation of $\Delta_2\{\dot{\Omega}, \dot{\omega}, \dot{M}\}$ to the power e^{2k} in the eccentricity, $10 \times 3 \times 2k$ terms must be evaluated using (3.72) in (3.38) to obtain $\dot{s}_{i20pq}$ and $\Delta_1 s_{j20pq}$. Then (V_{20} not being a function of Ω), $5 \times 6 \times 2k \times 3$ differentiations must be made to obtain the $\partial \dot{s}_{i20pq}/\partial s_j$; and finally, a like number of multiplications must be performed.

Some saving of effort is possible if we express V_{20} in the closed form of (3.61), evaluating $F_{20p}(i)$ by (3.62),

$$V_{20} = C_{20}\frac{\mu}{r}\left(\frac{a_e}{r}\right)^2\left[\left(\frac{3}{4}\sin^2 i - \frac{1}{2}\right) - \frac{3}{4}\sin^2 i \cos 2(\omega + f)\right], \tag{3.82}$$

using the partial derivatives of elliptic motion derivable from (3.9)–(3.19), and postponing the averaging with respect to time until the end. This method has been used in developing some theories. However, the more satisfying analyses of the close satellite problem have gone back to a considerably earlier point than the Lagrangian equations (3.38) in an attempt to obtain a clearer insight or a more accurate or more efficient solution. Some of the best solutions have expressed the potential in ellipsoidal harmonics; other solutions have used elements differing appreciably from the Keplerian in order to attain a formulation suitable for numerical iteration.

The solution we shall outline is that developed from the Delaunay equations (3.47). To abbreviate the equations, we express L, G, H as p_1, p_2, p_3 and M, ω, Ω as q_1, q_2, q_3 (not to be confused with the vector $\mathbf{q}$ in the orbital plane). Equation (3.47) then becomes

$$\dot{p}_i = \partial F/\partial q_i, \; \dot{q}_i = -\partial F/\partial p_i. \tag{3.83}$$

Since, from (3.83),

$$\frac{\partial F}{\partial p_i}\cdot\frac{dp_i}{dt} + \frac{\partial F}{\partial q_i}\cdot\frac{dq_i}{dt} = 0, \tag{3.84}$$

we have

$$\frac{dF}{dt} = \frac{\partial F}{\partial t}. \tag{3.85}$$

Here the explicit derivative $\partial F/\partial t$ signifies, in the problem posed by the disturbing potential (3.70),

$$\frac{\partial F}{\partial t} = \frac{\partial F}{\partial \theta}\dot{\theta}. \tag{3.86}$$

$\partial F/\partial t$ is thus purely periodic. Hence, if the solution is known for the motion expressed by the canonical set ($\mathbf{p}'$, $\mathbf{q}'$) with a constant force function F' close to F, the solution for ($\mathbf{p}$, $\mathbf{q}$) can be found as that for ($\mathbf{p}'$, $\mathbf{q}'$) plus a Taylor series over the difference ($\mathbf{p} - \mathbf{p}'$, $\mathbf{q} - \mathbf{q}'$). Several such transformations could be made in succession; the most general such succession for a close earth satellite would be

$$\begin{aligned} F(L, G, H, l, g, h, t) &\rightarrow F'(L', G', H', -, g', h', -) \\ &\rightarrow F''(L'', G'', H'', -, -, h'', -) \\ &\rightarrow F'''(L''', G''', H''', -, -, -, -). \end{aligned} \tag{3.87}$$

The last transformation is obviously solvable as

$$\begin{aligned} p_i''' &= \text{const}_i, \\ q_i''' &= q_{i0}''' + \text{const}_i(t - t_0). \end{aligned} \tag{3.88}$$

In the case of the disturbing potential V_{20} of (3.72) or (3.82), the last transformation in (3.87) is unnecessary since the potential does not contain the nodal longitude h.

The fact that the motions $(\mathbf{p}, \mathbf{q})$ and $(\mathbf{p}', \mathbf{q}')$ are derived from single scalars, F and F', respectively, suggests that the Taylor series transformation could also be expressed in terms of a single scalar. We start with two scalars,

$$x = \Sigma_i p_i \dot{q}_i + F,$$
$$y' = -\Sigma_i \dot{p}_i' q_i' + F', \tag{3.89}$$

We wish to prove that between fixed points $(\mathbf{p}, \mathbf{q})$ at times t_1 and t_2, variations of the integrals of x and y' will be zero for small variations $(\delta\mathbf{p}, \delta\mathbf{q})$ of the path. Thus we have

$$\delta\int_{t_1}^{t_2} (\Sigma_i p_i \dot{q}_i + F)\, dt = 0,$$
$$\delta\int_{t_1}^{t_2} (-\Sigma_i \dot{p}_i' q_i' + F')\, dt = 0, \tag{3.90}$$

known as Hamilton's principle.

Because the variable x is a function of p_i, q_i, and $\dot{q}_i$, we can write the first integral as

$$\int_{t_1}^{t_2} \left(\frac{\partial x}{\partial p_i}\,\delta p_i + \frac{\partial x}{\partial q_i}\,\delta q_i + \frac{\partial x}{\partial \dot{q}_i}\,\delta \dot{q}_i \right) dt. \tag{3.91}$$

From (3.89) and (3.83) we have

$$\frac{\partial x}{\partial p_i} = \dot{q}_i + \frac{\partial F}{\partial p_i} = 0, \tag{3.92}$$

which eliminates the first term of the integrand in (3.91). For the second term, from (3.89) we have

$$\frac{\partial x}{\partial q_i} = \frac{\partial F}{\partial q_i}. \tag{3.93}$$

In the third term, we replace $\delta\dot{q}_i$ by $d(\delta q_i)/dt$, and apply integration by parts. Thus

$$\int_{t_1}^{t_2} \frac{\partial x}{\partial \dot{q}_i} \frac{d}{dt}(\delta q_i)\, dt = \left(\frac{\partial x}{\partial \dot{q}_i}\,\delta q_i \right)_{t_1}^{t_2} - \int_{t_1}^{t_2} \frac{d}{dt}\left(\frac{\partial x}{\partial \dot{q}_i} \right) \delta q_i\, dt. \tag{3.94}$$

The first term vanishes because the variations δq_i are zero at the end points. From (3.89), $\partial x/\partial \dot{q}_i$ is p_i; thus between (3.93) and (3.94) we have, remaining as the integrand,

$$\left(\frac{\partial F}{\partial q_i} - \dot{p}_i\right)\delta q_i = 0, \tag{3.95}$$

which is zero by (3.83). The second equation in (3.90) can be proved similarly.

Hamilton's principle (3.90) indicates that the indefinite integrals are continuous, and hence that their difference is expressible by a single function S. Therefore the integrands differ at most by the total time derivative of S:

$$\int_{t_1}^{t_2} \frac{dS}{dt}\, dt = S(t_2) - S(t_1). \tag{3.96}$$

S is a function of both the $(\mathbf{p},\ \mathbf{q})$ and $(\mathbf{p}',\ \mathbf{q}')$ variables; however, if $(\mathbf{p},\ \mathbf{q})$ is to be expressed as $(\mathbf{p}',\ \mathbf{q}')$ plus a Taylor series development over the difference, then only one half of the twelve variables $(\mathbf{p},\ \mathbf{q},\ \mathbf{p}',\ \mathbf{q}')$ are independent, and S can be expressed in terms of any two of the four sets. The expression most commonly used is

$$S = S(\mathbf{q},\ \mathbf{p}'). \tag{3.97}$$

Then from (3.90), (3.96), and (3.97), summing over repeated subscripts,

$$\begin{aligned} \frac{dS}{dt} &= \frac{\partial S}{\partial q_i}\dot{q}_i + \frac{\partial S}{\partial p_i'}\dot{p}_i' + \frac{\partial S}{\partial t} \\ &= p_i\dot{q}_i + F + \dot{p}_i'q_i' - F'. \end{aligned} \tag{3.98}$$

Because q_i and p_i' are independent, for both values of dS/dt in (3.98) the coefficients of $\dot{q}_i$ and $\dot{p}_i'$, respectively, must be equal. Thus

$$p_i = \frac{\partial S}{\partial q_i}, \qquad q_i' = \frac{\partial S}{\partial p_i'}, \tag{3.99}$$

which leaves

$$F' = F - \frac{\partial S}{\partial t}. \tag{3.100}$$

S is called a determining function or generating function, and its use to effect a transformation $(\mathbf{p},\ \mathbf{q}) \to (\mathbf{p}',\ \mathbf{q}')$ in accordance with (3.87) is known as Von Zeipel's method. In applying this method, S, F', and F are all developed in series in accordance with the magnitude of the terms involved.

Thus

$$S = S_0 + S_1 + S_2 + \cdots, \tag{3.101}$$

$$F_0'(\mathbf{p}') + F_1'(\mathbf{p}', \mathbf{q}') + F_2'(\mathbf{p}', \mathbf{q}') + \cdots$$
$$= F_0(\mathbf{p}) + F_1(\mathbf{p}, \mathbf{q}) + F_2(\mathbf{p}, \mathbf{q}) + \cdots - \int \frac{\partial F}{\partial t}\, dt. \tag{3.102}$$

S_0 must have a form such that $p_i = p_i'$, $q_i = q_i'$, in the unperturbed case; that is,

$$S_0 = q_i p_i'. \tag{3.103}$$

Substituting (3.103) in (3.101), differentiating (3.101) with respect to q_i, p_i', and substituting the result in (3.99), we have

$$\begin{aligned} p_i &= p_i' + \frac{\partial S_1}{\partial q_i} + \frac{\partial S_2}{\partial q_i} + \cdots, \\ q_i &= q_i' - \frac{\partial S_1}{\partial p_i'} - \frac{\partial S_2}{\partial p_i'} - \cdots. \end{aligned} \tag{3.104}$$

Developing F in (3.100) in Taylor series of $(p_i - p_i')$ and $(q_i - q_i')$ and substituting for $(p_i - p_i')$ and $(q_i - q_i')$ from (3.104), we get

$$\left[\begin{aligned} F - \frac{\partial S}{\partial t} &= F_0(\mathbf{p}') \\ &+ \frac{\partial F_0}{\partial p_i'} \cdot \frac{\partial S_1}{\partial q_i} + F_1(\mathbf{p}', \mathbf{q}') - \frac{\partial S_1}{\partial t} \\ &+ \frac{\partial F_0}{\partial p_i'}\frac{\partial S_2}{\partial q_i} + \frac{1}{2}\frac{\partial^2 F_0}{\partial p_i' \partial p_j'}\left(\frac{\partial S_1}{\partial q_i}\right)\left(\frac{\partial S_1}{\partial q_j}\right) \\ &+ \frac{\partial F_1}{\partial p_i'}\frac{\partial S_1}{\partial q_i} + F_2(\mathbf{p}', \mathbf{q}') - \frac{\partial S_2}{\partial t} + \cdots \end{aligned}\right] = \left[\begin{aligned} &F_0' \\ &+F_1' \\ &+F_2' \\ &+\frac{\partial F_1'}{\partial q_i'}\frac{\partial S_1}{\partial p_i'}. \end{aligned}\right] \tag{3.105}$$

We then equate terms of equal magnitude in F and F'. These equations determine the terms in S, because F' will generally contain one less variable than F and hence any terms in F dependent on this omitted variable must be accounted for by S.

For the problem of interest to geodesy [using (3.39) and (3.46)],

$$\begin{aligned} F_0 &= \mu^2/2L^2, \\ F_1 &= V_{20}, \\ F_2 &= V_{lm \neq 00,20}. \end{aligned} \tag{3.106}$$

Let us consider the problem in which V_{20} is expressed by (3.72) and F_2 is zero. Then from the first line of (3.105),

$$F_0' = \mu^2/2L'^2. \tag{3.107}$$

On the second line of (3.105), we choose F_1' to absorb all terms not functions of M, or q_1. Thus

$$F_1' = \frac{\mu C_{20}}{a'}\left(\frac{a_e}{a'}\right)^2 F_{201}(i')G_{210}(e') = V'_{2010}, \tag{3.108}$$

since from (3.66), $G_{20(-2)}$ and G_{222} are zero. Then the remainder of the second line can be used to solve for $\partial S_1/\partial q_1$, since F_0 [from (3.106)] is a function of p_1 only:

$$\frac{\partial S_1}{\partial M} = \frac{\partial S_1}{\partial q_1} = \left(\frac{L'^3}{\mu^2}\right)\frac{\mu C_{20}}{a'}\left(\frac{a_e}{a'}\right)^2 \times \sum_{pq\neq 10} F'_{20p}G'_{2pq}\cos\left[(2-2p)\omega + (2-2p+q)M\right], \tag{3.109}$$

and

$$S_1 = \left(\frac{L'^3}{\mu^2}\right)\frac{\mu C_{20}}{a'}\left(\frac{a_e}{a'}\right)^2 \sum_{pq\neq 10} \frac{F'_{20p}G'_{2pq}}{(2-2p+q)}\sin\left[(2-2p)\omega + (2-2p+q)M\right]. \tag{3.110}$$

Derivatives of S_1 are then used in (3.104) to obtain the short period variations in the elements that will agree with those given by (3.76). In taking derivatives with respect to L', G', and H', it is convenient to use the partial derivatives of the Keplerian elements with respect to the Delaunay elements. The derivatives that are nonzero are

$$\begin{aligned} &\partial a/\partial L = 2L/\mu, \\ &\partial e/\partial L = G^2/L^3 e, \qquad \partial e/\partial G = -G/L^2 e, \\ &\partial i/\partial G = H/G^2 \sin i, \qquad \partial i/\partial H = -1/G \sin i. \end{aligned} \tag{3.111}$$

Proceeding to the third and fourth lines of (3.105), we are principally interested in the terms on the left arising from $(\partial S_1/\partial q_1)^2$ and $(\partial F_1/\partial p_i')\cdot(\partial S_1/\partial q_i)$ that will not have a sinusoidal factor because each of the cosine terms in F_1 and $\partial S_1/\partial q_i$ has been multiplied by itself. As a consequence, there are contributions to the secular motions $\dot{M}$, $\dot{\omega}$, $\dot{\Omega}$ through F_2' of coefficient C_{20}^2.

On the next transformation from F' to F'', terms containing the argument of perigee ω' are removed. In this transformation, the first two lines of (3.105) merely yield that F_0'' and F_1'' are equal to F_0' and F_1', respectively, since there are no periodic terms involved. On the third line both terms are zero, since S' does not contain M; but on the fourth line, there are two nonzero terms:

$$\frac{\partial F_1'}{\partial G''} \cdot \frac{\partial S_1'}{\partial \omega'} + F_2' = F_2'' \tag{3.112}$$

$\partial F_1'/\partial G''$ is, from (3.83), equal to $\dot{\omega}$, which by (3.74), contains a factor $(1 - 5\cos^2 i)$. Hence in solving for $\partial S_1'/\partial\omega'$ we obtain a factor $(1 - 5\cos^2 i)^{-1}$, and the solution is not valid in the vicinity of an inclination $\cos^{-1}\sqrt{1/5}$, which must be treated as a case of resonance.

A clear and detailed exposition of the application of the Von Zeipel method with F_1 as in (3.106) is given by Brouwer (1959); see also Brouwer and Clemence (1961, pp. 562–573). The final secular and long-period terms to order J_2^2 are

$$\begin{aligned}
M' &= M_0'' + n_0 t\{1 + \tfrac{3}{2}\gamma_2'\eta(-1 + 3\theta^2) + \tfrac{3}{32}\gamma_2'^2\eta[-15 + 16\eta + 25\eta^2 \\
&\quad + (30 - 96\eta - 90\eta^2)\theta^2 + (105 + 144\eta + 25\eta^2)\theta^4]\} \\
&\quad + \tfrac{1}{8}\gamma_2'\eta^3[1 - 11\theta^2 - 40\theta^4(1 - 5\theta^2)^{-1}]\sin 2\omega'', \\
\omega' &= \omega_0'' + n_0 t\{\tfrac{3}{2}\gamma_2'(-1 + 5\theta^2) + \tfrac{3}{32}\gamma_2'^2[-35 + 24\eta + 25\eta^2 \\
&\quad + (90 - 192\eta - 126\eta^2)\theta^2 + (385 + 360\eta + 45\eta^2)\theta^4]\} \\
&\quad - \tfrac{1}{16}\gamma_2'[(2 + e''^2) - 11(2 + 3e''^2)\theta^2 - 40(2 + 5e''^2)\theta^4(1 - 5\theta^2)^{-1} \\
&\quad - 40e''^2\theta^6(1 - 5\theta^2)^{-2}]\sin 2\omega'', \\
\Omega' &= \Omega_0'' + n_0 t\{-3\gamma_2'\theta + \tfrac{3}{8}\gamma_2'^2[(-5 + 12\eta + 9\eta^2)\theta + (-35 - 36\eta - 5\eta^2)\theta^3]\} \\
&\quad - \tfrac{1}{8}\gamma_2' e''^2\theta[11 + 80\theta^2(1 - 5\theta^2)^{-1} + 200\theta^4(1 - 5\theta^2)^{-2}]\sin 2\omega'', \\
e' &= e'' + \tfrac{1}{8}\gamma_2' e''\eta^2[1 - 11\theta^2 - 40\theta^4(1 - 5\theta^2)^{-1}]\cos 2\omega'', \\
i' &= i'' - \tfrac{1}{8}\gamma_2' e''^2\cot i''[1 - 11\theta^2 - 40\theta^4(1 - 5\theta^2)^{-1}]\cos 2\omega'',
\end{aligned} \tag{3.113}$$

where $\eta = (1 - e''^2)^{1/2}$, $\theta = \cos i''$, $\gamma_2' = a_e^2 J_2/(2a''^2\eta^4)$, and $n_0 = \mu^{1/2}a''^{-3/2}$. The constants of integration defined by Brouwer's theory are the mean elements: namely the constant parts a'', e'', i'' and the secularly changing parts at epoch M_0'', ω_0'', Ω_0'', that are consistent with (3.88).

In addition to the J_2^2 terms, there are two other types of second-order terms that may be of significance:

1. If short period variations of the mean anomaly M are being taken into account, then variations in M resulting from changes in the mean

motion n arising from perturbations of the semimajor axis should be included. From (3.80),

$$\Delta_2 M = \int \frac{\partial n}{\partial a} \Delta_1 a \, dt. \tag{3.114}$$

Using (3.76) for $\Delta_1 a$ and $-3n/2a$ for $\partial n/\partial a$,

$$\Delta_2 M_{lmpq} = -\frac{3\mu a_e^l F_{lmp} G_{lpq} \bar{S}_{lmpq}(l - 2p + q)}{a^{l+3}[(l - 2p)\dot{\omega} + (l - 2p + q)\dot{M} + m(\dot{\Omega} - \dot{\theta})]^2}. \tag{3.115}$$

2. If l is odd and m is zero, the interactions between the period perturbations $\Delta_1 e_{lop(2p-l)}$, $\Delta_1 i_{lop(2p-l)}$, and the secular motions $\dot{\Omega}_{2010}$ and $\dot{\omega}_{2010}$ can have perceptible effect on the periodic perturbations $\Delta\Omega_{lop(2p-l)}$, $\Delta\omega_{lop(2p-l)}$, and $\Delta M_{lop(2p-l)}$. Using (3.80) again and abbreviating $lop(2p - l)$ as r, we get

$$\begin{aligned} \Delta_2 \Omega_r &= \frac{\partial \dot{\Omega}}{\partial e} \int \Delta_1 e_r \, dt + \frac{\partial \dot{\Omega}}{\partial i} \int \Delta_1 i_r \, dt \\ &= \frac{3C_{20} n}{2(1 - e^2)^2} \left(\frac{a_e}{a}\right)^2 \left(\frac{4e \cos i}{(1 - e^2)} \int \Delta_1 e_r \, dt - \sin i \int \Delta_1 i_r \, dt\right). \end{aligned} \tag{3.116}$$

In (3.116), (3.74) has been applied. Similar expressions can be obtained for $\Delta_2 \omega_r$ and $\Delta_2 M_r$.

3.6. Resonance

The methods described thus far in this chapter fail for two categories of orbits:

1. Orbits for which the eccentricity e, or inclination i, is so close to zero that nodal longitude Ω, perigee argument ω, or anomaly M, may lose their definition and have absurdly large perturbations that virtually cancel each other in calculation of position or velocity.

2. Orbits for which the secular rates of the arguments in some terms of the disturbing function, (3.70), may be so close to zero that their periodic variation is more significant; that is, we have a resonant situation in which there may be libration rather than secular motion:

$$(l - 2p)\dot{\omega} + (l - 2p + q)\dot{M} + m(\dot{\Omega} - \dot{\theta}) \approx 0. \tag{3.117}$$

The difficulties in category (1)—near-zero eccentricity or inclination—are somewhat artificial, in that they do not involve any physical phenomenon, but are a matter of definition of coordinates, solvable by changing the orbital

elements. Two types of changes made to canonical elements ($\mathbf{p}$, $\mathbf{q}$) such as the Delaunay elements (3.43), (3.45), and (3.46) are
for any two pairs i, j,

$$\begin{aligned} p_i' &= p_i, & q_i' &= q_i \pm q_j, \\ p_j' &= p_j \mp p_i, & q_j' &= q_j, \end{aligned} \tag{3.118}$$

and for any i,

$$p_i' = \sqrt{2p_i}\cos q_i, \qquad q_i' = \sqrt{2p_i}\sin q_i.$$

Such transformations can be applied to any pair, or any two pairs, in succession.

The difficulties in category (2) are quite real; there is an actual change in the behavior of the orbit. The case that has been treated most extensively is that of the "critical inclination" $\cos^{-1}\sqrt{1/5}$, for which, as mentioned in connection with (3.112),

$$\dot{\omega} \approx 0. \tag{3.119}$$

A case of greater interest, both because it has actually occurred in connection with satellite communication projects and because it can yield information about the gravitational field, is

$$\dot{\omega} + \dot{M} + \dot{\Omega} - \dot{\theta} \approx 0. \tag{3.120}$$

A satellite with the mean secular rate (3.120) will resonate with any spherical harmonic C_{lm}, S_{lm} that gives rise to terms with an integral multiple of $(\omega + M + \Omega - \theta)$ as an argument. A rough calculation setting $\dot{\theta}$ for n and using Kepler's law (3.20) indicates that the semimajor axis for the case (3.120) to occur is about 6.6 earth radii. Hence, because of the $(a_e/a)^l$ term in the disturbing function (3.70), the dominant effect will be by those terms for which l is small—in particular, C_{22}, S_{22}.

Since $l - 2p$ must equal m to obtain an integral multiple of $(\omega + M + \Omega - \theta)$ as an argument in the disturbing function (3.70), $l - m$ must be even, and p must be $(l - m)/2$. Also q must be zero. Let

$$Q_{lm} = \frac{\mu}{a}\sqrt{C_{lm}^2 + S_{lm}^2}\left[\frac{a_e}{a}\right]^l F_{lm,(l-m)/2}(i)G_{l,(l-m)/2,0}(e) \tag{3.121}$$

and

$$\lambda_{lm} = \frac{1}{m}\tan^{-1}\left(\frac{S_{lm}}{C_{lm}}\right). \tag{3.122}$$

From (3.70) the resonating disturbing function can be written

$$R = \sum_{(l-m)\text{even}} Q_{lm} \cos m(\omega + M + \Omega - \lambda_{lm} - \theta). \qquad (3.123)$$

Since the mean anomaly M appears in this disturbing function, there will be a perturbation of the semimajor axis, from the first of the equations of motion (3.38):

$$\dot{a} = -\frac{2}{na} \sum_{(l-m)\text{even}} mQ_{lm} \sin m(\omega + M + \Omega - \lambda_{lm} - \theta). \qquad (3.124)$$

This change in the semimajor axis in turn causes an acceleration in the earth-referred longitude of the satellite. Let

$$\lambda_A = \omega + M + \Omega - \theta, \qquad (3.125)$$

where we have added the subscript A to indicate that this is the "broken-legged" astronomical longitude, not the conventional geodetic longitude measured along the equator. Using Kepler's law (3.20), we get

$$\ddot{\lambda}_A = \ddot{M} = \frac{\partial n}{\partial a} \dot{a} = \frac{3}{a^2} \sum_{(l-m)\text{even}} mQ_{lm} \sin m(\lambda_A - \lambda_{lm}). \qquad (3.126)$$

The double integration of (3.126) in a manner similar to (3.115) would lead to a large $\Delta\lambda_A$ because of the small divisor $(n - \dot{\theta})^2$. In addition, there would be terms with divisor $(n - \dot{\theta})$ obtained by using the disturbing function (3.123) in an integration of the equations of motion (3.38) in the manner of (3.76). However, both (3.76) and (3.115) are derived under the assumption that a valid first approximation of the orbit is a secularly precessing ellipse. Such approximations usually break down in the vicinity of a resonance, so (3.126) must be examined directly.

By multiplying (3.126) by $2\dot{\lambda}_A$ we obtain, as a first integral with constant of integration K,

$$(\dot{\lambda}_A)^2 = K - \frac{6}{a^2} \sum_{(l-m)\text{even}} Q_{lm} \cos m(\lambda_A - \lambda_{lm}). \qquad (3.127)$$

Let us define the constant K as a combination of the initial longitude λ_{A0} and rate $\dot{\lambda}_{A0}$:

$$K = (\dot{\lambda}_{A0})^2 + \frac{6}{a^2} \sum_{(l-m)\text{even}} Q_{lm} \cos m(\lambda_{A0} - \lambda_{lm}). \qquad (3.128)$$

Since $(\dot{\lambda}_A)^2$ must be positive, from (3.127) we have the condition

$$K \geq \frac{6}{a^2} \sum_{(l-m)\text{even}} Q_{lm} \cos m(\lambda_A - \lambda_{lm}). \tag{3.129}$$

If the constant K is small enough, the condition (3.129) will prevent the longitude λ_A from going through a full cycle from 0 to 2π. A small K is most likely to result if λ_{A0} in (3.128) is such as to make the dominant term $Q_{22} \cos 2(\lambda_{A0} - \lambda_{22})$ near $-Q_{22}$: that is, if λ_{A0} is near $\lambda_{22} \pm \pi/2$, which is the minor axis of the equatorial ellipse. On referring longitudes to this minor axis by making the substitution

$$\psi = \lambda_A - \lambda_{22} - \pi/2 \tag{3.130}$$

and taking only the $lm = 22$ term of (3.127), we get

$$\begin{aligned}(\dot{\psi})^2 &= K + \frac{6}{a^2} Q_{22} \cos 2\psi \\ &= K + \frac{6}{a^2} Q_{22}[1 - 2\sin^2 \psi].\end{aligned} \tag{3.131}$$

The maximum departure ψ_m from the minor axis will thus correspond to zero $(\dot{\psi})^2$,

$$\sin \psi_m = \left[\frac{Ka^2}{12Q_{22}} + \frac{1}{2}\right]^{1/2}. \tag{3.132}$$

From (3.131) we have

$$\begin{aligned}\frac{d\psi}{dt} &= \left[K + \frac{6Q_{22}}{a^2} - \frac{12Q_{22}}{a^2} \sin^2 \psi\right]^{1/2} \\ &= \left[\frac{12Q_{22}}{a^2}(\sin^2 \psi_m - \sin^2 \psi)\right]^{1/2}.\end{aligned} \tag{3.133}$$

On defining

$$k^2 = 1/\sin^2 \psi_m, \tag{3.134}$$

shifting dt and functions of ψ to opposite sides of the equation, and integrating, (3.133) becomes

$$t = \frac{ak}{(12Q_{22})^{1/2}} \int_0^\psi \frac{d\psi}{[1 - k^2 \sin^2 \psi]^{1/2}} = \frac{ak}{(12Q_{22})^{1/2}} F(k, \psi). \tag{3.135}$$

F is the elliptic integral of the first kind, which for cases where $k < 1$ are discussed at length in textbooks such as Hancock (1917). For the interesting cases in the present problem that involve libration, $k > 1$. In order to evaluate the resulting integral, we use the transformation

$$F(k, \psi) = \frac{1}{k} F\left[\frac{1}{k}, \sin^{-1}(k \sin \psi)\right]. \qquad (3.136)$$

Therefore a complete cycle of the elliptic integral F in (3.135) is $(4/k)F(1/k, \pi/2)$, and the time T for a complete period will be

$$T = \frac{2a}{(3Q_{22})^{1/2}} F\left(\frac{1}{k}, \frac{\pi}{2}\right). \qquad (3.137)$$

If k is greater than unity, and the initial longitude λ_0 is close enough to the stable point $\lambda_{22} \pm \pi/2$, $\sin \psi$ must vary between the limits $\pm(1/k)$; thus, the satellite will librate about the minimum longitude $\lambda_{22} + \pi/2$. If k is less than unity or if the initial longitude λ_0 is outside the limits of libration, ψ may have any value: that is, the satellite will drift all the way around the earth. Considering that the satellite mean motion n must be greater than the earth's rotation rate $\dot{\theta}$ inside the zone of possible resonance, and less outside this zone, we can draw a schematic picture of the areas of libration in an earth-fixed reference frame, as shown in Figure 5.

An analysis of resonance can also be made similar to the Von Zeipel method of Equations (3.83)–(3.105). In order to remove the explicit appearance of time through the angle θ in the disturbing function (3.123), the third angle variable q_3 can be made:

$$h = \Omega - \theta. \qquad (3.138)$$

Then in order for the equations of motion (3.83) to apply, $\dot{\theta}H$ must be added to the force function F. By taking only the $lm = 22$ term from the disturbing function (3.123), replacing $\cos 2(\omega + M + h - \lambda_{22})$ by $2\cos^2(\omega + M + h - \lambda_{22}) - 1$, and breaking down F by magnitude in accordance with (3.102), we get

$$\begin{aligned} F_0 &= \mu^2/2L^2 + \dot{\theta}H, \\ F_1 &= 2\, Q_{22} \cos^2(\omega + M + h - \lambda_{22}), \\ F_2 &= -Q_{22}. \end{aligned} \qquad (3.139)$$

If we develop the Von Zeipel transformation as in (3.105), we get

$$\left.\begin{matrix} F_0 \\ + \dfrac{\partial F_0}{\partial L'}\dfrac{\partial S_1}{\partial M} + \dfrac{\partial F_0}{\partial H'}\dfrac{\partial S_1}{\partial h} + \dfrac{1}{2}\dfrac{\partial^2 F_0}{\partial L'^2}\left(\dfrac{\partial S_1}{\partial M}\right)^2 + F_1 \\ + \dfrac{\partial F_0}{\partial L'}\dfrac{\partial S_2}{\partial M} + \dfrac{\partial F_0}{\partial H'}\dfrac{\partial S_2}{\partial h} \qquad\qquad + F_2 \end{matrix}\right] = \begin{matrix} F_0' \\ \\ +F_2'. \end{matrix} \qquad (3.140)$$

Here F_1' does not appear on the right of (3.140) since the corresponding part of F is entirely periodic, and hence must be absorbed by S_1. $\partial S_1/\partial M$ and $\partial S_1/\partial h$ are equal, since M and h appear only as a sum $(M + h)$. The phenomenon of resonance thus appears in the form of a very small coefficient to this derivative since, from (3.139), (3.46), and (3.20), $\partial F_0/\partial L'$ is the negative

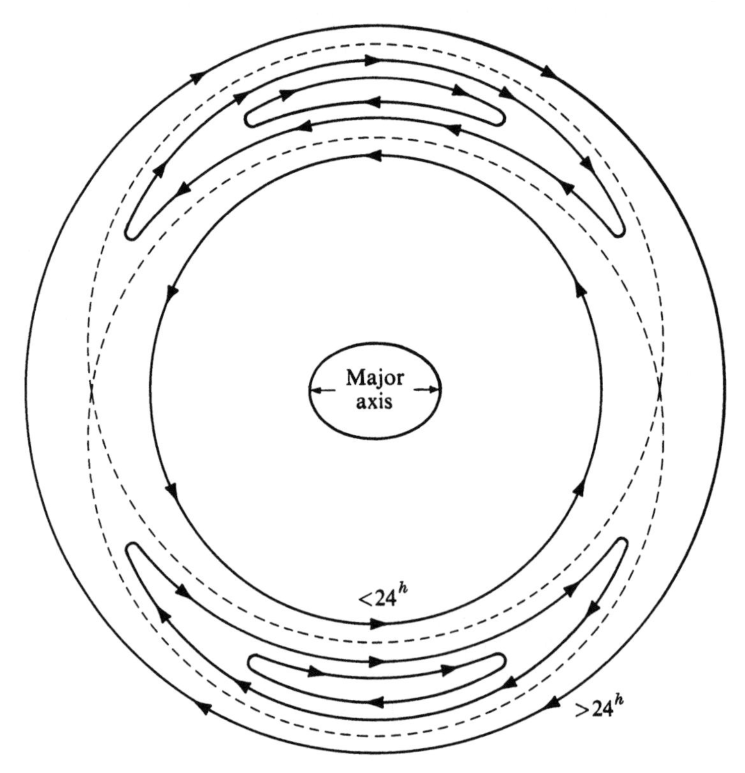

FIGURE 5. *Near 24-hour orbit paths in an earth-fixed reference frame.*

of the mean motion, $-n'$. It is for this reason that the change from (3.105) was made in (3.140) of shifting the $(\partial S_1/\partial M)^2$ term to the second line. On substituting ψ as defined by (3.130) and (3.125), the second line becomes

$$(\dot{\theta} - n')\frac{\partial S_1}{\partial \psi} + \frac{3n'}{2L'}\left(\frac{\partial S_1}{\partial \psi}\right)^2 + 2Q_{22}\sin^2\psi = 0. \tag{3.141}$$

Solving the quadratic equation (3.141) for $\partial S_1/\partial \psi$ and integrating, we get

$$S_1 = -(\dot{\theta} - n')(\psi \pm E)L'/3n', \tag{3.142}$$

where

$$E = E(k, \psi) = \int_0^{\psi} (1 - k^2 \sin^2 \psi)^{1/2} \, d\psi, \tag{3.143}$$

the elliptic integral of the second kind, in which the modulus is

$$k^2 = \frac{12Q_{22}n'}{(\dot{\theta} - n')^2 L'}. \tag{3.144}$$

For the longitude λ_A with respect to an earth-fixed point, (3.104) can be applied, taking derivatives of the determining function with respect to L', G', H'. After considerable algebra and eliminating minor terms, a result similar to (3.137) is obtained.

The method developed in (3.121)–(3.137) is most similar to that of Allan (1963). Morando (1963) has applied the Von Zeipel transformation to the problem, as outlined by (3.138)–(3.144).

Another case of possibly significant approach to resonance in accordance with (3.117) that has been suggested is

$$\dot{\omega} + \dot{M} + m(\dot{\Omega} - \dot{\theta}) \approx 0. \tag{3.145}$$

For moderate eccentricities $(l - 2p)$, and hence l, must be odd. The semi-major axis must satisfy the condition

$$a = \left[\frac{\mu}{n^2}\right]^{1/3} = \left[\frac{\mu}{(m\dot{\theta})^2}\right]^{1/3} \approx \left[\frac{17}{m}\right]^{2/3} a_e > a_e. \tag{3.146}$$

Since $l \geq m$, and since $a^{-(l+3/2)}$ appears in the perturbation equations (3.76), higher values of m would appear to be more effective, despite the drop off in magnitude of the coefficients with increase in l. Then, for a particular

$$m \approx 17\left[\frac{a_e}{a}\right]^{3/2}, \tag{3.147}$$

the disturbing function after (3.70) will be

$$R = \frac{\mu a_e^{m+1-k}}{a^{m+2-k}} \sum_{j=0}^{\infty} \left(\frac{a_e}{a}\right)^{2j} F_{lmp}(i) G_{lp0}(e) S_{lmp0}, \tag{3.148}$$

where

$$\begin{aligned} k &= m (\text{mod } 2), \\ l &= m + 2j + 1 - k, \\ p &= (m - k)/2 + j. \end{aligned} \tag{3.149}$$

For example, for $a \approx 1.196a_e$, we have $m = 13$; $l = 13, 15, 17, \cdots$; and $p = 6, 7, 8, \cdots$.

Since $\dot{\theta}$ is one cycle per day, for any satellite there will be at least one term of a period of two days or more. Hence, the second-order term in the anomaly $\Delta_2 M$ from (3.115) becomes dominant because the rate is squared in the denominator. Then for the perturbation along track, from (3.115) and (3.76),

$$\begin{aligned}\Delta\lambda &= \Delta\Omega \cos i + \Delta\omega + \Delta_1 M + \Delta_2 M \\ &= \frac{\mu a_e^{m+1-k}}{a^{m+4-k}[\dot{\omega} + \dot{M} + m(\dot{\Omega} - \dot{\theta})]} \sum_{j=0}^{\infty} \left(\frac{a_e}{a}\right)^{2j} F_{lmp} \bar{S}_{lmp0} \\ &\quad \times \left\{\frac{1}{n}\left[H(e)\frac{\partial G_{lp0}}{\partial e} + 2(l+1)G_{lp0}\right] - \frac{3G_{lp0}}{[\dot{\omega} + \dot{M} + m(\dot{\Omega} - \dot{\theta})]}\right\}, \end{aligned} \tag{3.150}$$

where

$$H(e) = \frac{\sqrt{1-e^2} - (1-e^2)}{e} = \frac{e}{2} - \frac{3e^3}{8} - \frac{5e^5}{16} - \cdots. \tag{3.151}$$

Since they have opposite signs within the brackets, a rate $[\dot{\omega} + \dot{M} + m(\dot{\Omega} - \dot{\theta})]$ having the opposite sign from the mean motion n would yield the greatest effect.

3.7. Miscellaneous Effects

In addition to the perturbations caused by the variations of the earth's gravitational field, close satellite orbits will also be perturbed by the gravitational attractions of the sun and moon; the radiation pressure of the sun; and the drag of the atmosphere. These effects may be particularly significant in analyzing long period and secular variations to determine zonal harmonics. The gravitational perturbations resulting from the sun and moon can be developed analytically in a manner very similar to (3.76). Because the effect of the earth's shadow is most important in the radiation pressure perturbations and makes their analytical solution rather awkward, numerical integration or numerical harmonic analysis is normally applied.

It is even more complicated to calculate the orbital perturbations resulting from drag using a model of the upper atmosphere; therefore numerical methods are also indicated. However, in this case there is an additional difficulty in that even the best of atmospheric models are inadequate to account for much of the significant variation in density; thus it becomes pointless to make elaborate calculations of perturbations by the model. Instead, a few parameters expressing the principal drag perturbations are usually determined from the orbital tracking data itself, and some second-order effects calculated from these parameters.

At the altitudes of geodetically useful satellites, the force of drag can be expressed by

$$F_d = \frac{C_D}{2} A\rho v^2, \tag{3.152}$$

where C_D, about 2.4, depends on the shape of the satellite and the manner of reflection of the air molecules; A is the cross-sectional area; ρ is the atmospheric density; and v is the velocity of the satellite relative to the atmosphere. The drag force vector is always directed contrary to the velocity vector, and hence its effect is not "averaged out" by the rotation of the earth or of the orbit, as are the gravitational effects. Consequently, there is an energy loss which results in a contraction of the orbit and a speeding up of the satellite to counteract the increased gravitational pull, in accordance with Kepler's law, (3.20). This speeding up causes the drag perturbations to appear in the mean anomaly much more than in any other of the Keplerian elements, and hence the arbitrary parameters approximating the drag (plus, usually, radiation pressure) effects are usually coefficients M_i of a power series with respect to time of the mean anomaly,

$$\Delta M_d = \sum_{j=2}^{j\,\max} M_j(t - t_0)^j. \tag{3.153}$$

The series starts with M_2 because M_0 is one of the constants of integration and M_1 is indistinguishable from a_0 because of Kepler's law, (3.20). Equation (3.20) also gives the drag perturbations of the semimajor axis consequent to (3.153),

$$\Delta a_d = -\frac{2a^{5/2}}{3\mu^{1/2}} \sum_{j=2}^{j\,\max} jM_j(t - t_0)^{j-1}. \tag{3.154}$$

The atmospheric density decreases rapidly with altitude; so much so, that even for rather moderate eccentricities the drag can be considered almost as an impulse at perigee. Hence the energy loss can be related directly to the perigee radius r_p; from (3.25)

$$\Delta v^2 = 2\Delta T_p = \mu\,\Delta\left[\frac{2}{r_p} - \frac{1}{a}\right] = \mu\,\Delta\left[\frac{(1+e)}{a(1-e)}\right]. \tag{3.155}$$

Since $\Delta T_p < 0$ and $\Delta a < 0$, necessarily $\Delta e < 0$. To the first approximation r_p is constant, since there will not be change at right angles to an imposed force. On assuming r_p to be constant, we get

$$0 = \Delta a_d(1 - e) - a\,\Delta e_d,$$

$$\Delta e_d = \frac{\Delta a_d}{a}(1 - e)$$

$$= -\frac{2a^{3/2}}{3\mu^{1/2}}(1 - e)\sum_j jM_j(t - t_0)^{j-1}. \tag{3.156}$$

The rate of motion of the node and perigee caused by the oblateness C_{20}, or J_2, depends on the semimajor axis a and eccentricity e. Hence, there will be an acceleration of the node and perigee due to a change in a, e:

$$\Delta\Omega_d = \int \Delta\dot{\Omega}_d\,dt = \int \left[\frac{\partial\dot{\Omega}(C_{20})}{\partial a}\Delta a_d + \frac{\partial\dot{\Omega}(C_{20})}{\partial e}\Delta e_d\right] dt. \tag{3.157}$$

Evaluating $\dot{\Omega}(C_{20})$, $\dot{\omega}(C_{20})$ by (3.74), substituting from (3.154) and (3.156) for Δa_d, Δe_d, and integrating, we get

$$\Delta\Omega_d = \frac{\cos i}{2(1 - e^2)^2}\left[\frac{7 - e}{1 + e}\right] C_{20}\left(\frac{a_e}{a}\right)^2 \Delta M_d,$$

$$\Delta\omega_d = \frac{4 - 5\sin^2 i}{4(1 - e^2)^2}\left[\frac{7 - e}{1 + e}\right] C_{20}\left(\frac{a_e}{a}\right)^2 \Delta M_d. \tag{3.158}$$

In addition to the variations caused by atmospheric density expressed by (3.153), (3.154), (3.156), and (3.158), there are variations in the motion of the atmosphere that affect the orbit, since the velocity v in (3.145) is relative to the atmosphere. Such variations of motion appear most clearly in the inclination, which is unaffected by density variations. In particular, most orbits show secular decrease in the inclination, as would be expected if the atmosphere rotated with the earth, but at a rate fluctuating appreciably from that indicating a uniform rotation. Hence a treatment for the effects of variations in atmospheric motion starting from a polynomial for the inclination similar to (3.153)–(3.158) could be developed.

The treatment of drag is the most unsatisfactory part of satellite orbit dynamics. Short-term variations which are of significance in determining tesseral harmonics and station position shifts must be treated as a statistical problem, as discussed in Chapter 5. Long-term variations in the semimajor axis, eccentricity, and inclination must either be taken into account along the lines of (3.154) or averaged out carefully in using the changes in node $\Delta\Omega$ and perigee $\Delta\omega$ to determine zonal harmonics. For a detailed description of the dynamical theory of drag perturbations mainly to determine atmospheric properties, see the text by King-Hele (1964) and the review by Jacchia (1963).

3.8. Summary

The purpose of this chapter has been to develop the theory of close-satellite orbits as part of the geodetic environment—that is, to describe a particular phenomenon connected with the earth's gravity field. Hence the discussion has not been a complete overall treatment of the subject of close satellite orbits, but rather has emphasized those aspects of peculiar interest to geodesy, such as the development of the disturbing function for spherical harmonic variations in the gravitational field, and has neglected other aspects of lesser interest. It also has gone further into basic theory than is perhaps necessary for practical application, in order to improve understanding and to furnish a basis for more effective treatment of special problems such as 24-hour orbits.

REFERENCES

1. Allan, R. R. "Perturbations of a Geostationary Satellite:1. The Ellipticity of the Equator." *Planetary Space Sci.*, *11* (1963), pp. 1325–1334.

2. Brouwer, Dirk. "Solution of the Problem of Artificial Satellite Theory Without Drag." *Astron. J.*, *64* (1959), pp. 378–397.

3. Brouwer, Dirk, and G. M. Clemence. *Methods of Celestial Mechanics*. New York:Academic Press, Inc., 1961.

4. Cayley, A. "Tables of the Developments of Functions in the Theory of Elliptic Motion." *Mem. Roy. Astron. Soc.*, *29* (1861), pp. 191–306.

5. Conte, S. D. "The Computation of Satellite Orbit Trajectories." *Advan. Computers*, *3* (1962), pp. 1–76.

6. Cook, A. H. "The Contribution of Observations of Satellites to the Determination of the Earth's Gravitational Potential." *Space Sci. Rev.*, *2* (1963), pp. 355–437.

7. Hancock, H. *Elliptic Integrals*. New York:Dover Publications, Inc., 1917, republished, 1958.

8. Jacchia, L. G. "Variations in the Earth's Upper Atmosphere as Revealed by Satellite Drag." *Rev. Mod. Phys.*, *35* (1963), pp. 973–991.

9. Kaula, W. M. "Analysis of Gravitational and Geometric Aspects of Geodetic Utilization of Satellites." *Geophys. J.*, *5* (1961), pp. 104–133.

10. Kaula, W. M. "Gravitational and Other Perturbations of a Satellite Orbit." *Dynamics of Rockets and Satellites*. Amsterdam:North-Holland Publishing Company, 1965, pp. 179–215.

11. King-Hele, D. G. *Theory of Satellite Orbits in an Atmosphere*. London: Butterworth and Company, 1964.

12. Morando, B. "Orbites de Resonance des Satellites de 24 h." *Bull. Astron.*, *24* (1963), pp. 47–67.

13. Mueller, I. I. *Introduction to Satellite Geodesy*. New York: Frederick Ungar Publishing Company, Inc., 1964.

14. Tisserand, F. *Traité de Mécanique Celeste, Vol. 1: Perturbations des Planetes d'aprés la Methode de la Variation des Constantes Arbitraires.* Paris: Gauthier-Villars et Fils, 1889, republished, 1960.

4

GEOMETRY OF SATELLITE OBSERVATIONS

4.1. General

Thus far, there has been developed the theory of the motion of a particle in the earth's gravitational field, without reference to any observer. The only geometry introduced has been that necessary to the transformations of the potential and to the dynamics of the orbit. In this chapter we discuss the geometry pertaining to observation of the satellite; develop differential relationships for all quantities affecting observations; and consider variations in the reference frame, the medium through which observations are made, and so on.

4.2. Coordinate Transformations

In Chapter 2, we used a rectangular-coordinate system $\{u, v, w\}$ fixed in the earth. In geodesy, however, the ellipsoidal coordinates ϕ (latitude), λ (longitude), and h (altitude) are more commonly used. We are therefore interested in relating these ellipsoidal coordinates to the rectangular coordinates. Figure 6 shows a meridional section through the rotation ellipsoid. Substituting p^2 for $u^2 + v^2$, we have from (1.44) or (3.8)

$$\frac{p^2}{a^2} + \frac{w^2}{b^2} = 1, \tag{4.1}$$

where a is the equatorial and b is the polar semiaxis of the ellipsoid. Then from the figure and (4.1),

$$\tan \phi = -\frac{dp}{dw} = \frac{a^2}{b^2} \cdot \frac{w}{p} = \frac{w}{p(1 - e^2)}, \tag{4.2}$$

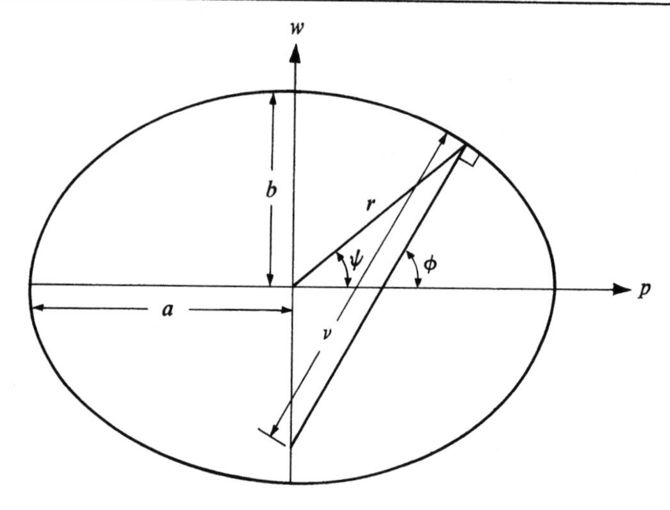

FIGURE 6. *Meridian ellipse.*

where e is the eccentricity defined in Figure 2. Also from Figure 6,

$$\cos \phi = p/\nu, \tag{4.3}$$

whence

$$\sin \phi = \frac{w}{\nu(1 - e^2)}. \tag{4.4}$$

Returning to the ellipsoid, and allowing for the altitude h, we have

$$\mathbf{u} = \begin{cases} u = (\nu + h) \cos \phi \cos \lambda, \\ v = (\nu + h) \cos \phi \sin \lambda, \\ w = [\nu(1 - e^2) + h] \sin \phi. \end{cases} \tag{4.5}$$

In order to define ν, use (4.3) and (4.1) again to get

$$\nu^2 \cos^2 \phi = p^2 = -\frac{a^2}{b^2} w^2 + a^2. \tag{4.6}$$

Then, from (4.4) and (4.6),

$$\nu^2(1 - \sin^2 \phi) = -\frac{1}{1 - e^2} \nu^2(1 - e^2)^2 \sin^2 \phi + a^2, \tag{4.7}$$

whence

$$\nu = \frac{a}{(1 - e^2 \sin^2 \phi)^{1/2}}. \tag{4.8}$$

Finally, to relate the eccentricity and the flattening, we have

$$\begin{aligned} e^2 &= \frac{a^2 - b^2}{a^2} = \left(\frac{a - b}{a}\right)^2 + 2\,\frac{ab - b^2}{a^2} \\ &= f^2 + 2(1 - f)f \\ &= 2f - f^2. \end{aligned} \tag{4.9}$$

In all the relationships developed so far, a coordinate origin at the center of the earth has been used. However, for use with observations from a station at geodetic location $\mathbf{u}_0 = (u_0, v_0, w_0)$, a translation from the center of the earth to the location is required. Using the subscript T to denote "topocentric" coordinates referred to such a point, we have

$$\mathbf{x}_T = \mathbf{R}_{xu}\mathbf{u}_T = \mathbf{R}_{xu}(\mathbf{u} - \mathbf{u}_0) \tag{4.10}$$

or

$$\mathbf{x}_T = \mathbf{R}_{xq}\mathbf{q} - \mathbf{R}_3(-\theta)\mathbf{u}_0, \tag{4.11}$$

where $\mathbf{R}_{xq}$ is defined by (2.32) and $\mathbf{R}_3(-\theta)$ by (2.24).

In order to obtain relative velocities, or rate-of-change of topocentric coordinates, the rotation of the earth, which causes the station to move with respect to inertial axes, must be taken into account as well as the motion of the satellite:

$$\dot{\mathbf{x}}_T = \mathbf{R}_{xq}\dot{\mathbf{q}} - \frac{\partial \mathbf{R}_{xu}}{\partial \theta}\,\mathbf{u}_0\dot{\theta}, \tag{4.12}$$

where $\dot{\mathbf{q}}$ is given by (3.24). Equation (4.12) applies regardless of whether or not the orbit is perturbed. Even though the Keplerian elements may all have nonzero rates of change, only the acceleration is affected, since the total velocity is represented by the $\dot{x}_i$ in (3.26), which is completely accounted for by the osculating Keplerian elements in a transformation such as that at end of Section 3.1. Alternatively, velocity can be referred to earth-fixed coordinates:

$$\dot{\mathbf{u}}_T = \mathbf{R}_{uq}\dot{\mathbf{q}} + \frac{\partial \mathbf{R}_{uq}}{\partial \theta}\,\mathbf{q}\dot{\theta}. \tag{4.13}$$

For the range r from a station to a satellite, by using the sum of squares of the rectangular coordinates, $x_1^2 + x_2^2 + x_3^2$, we have

$$r = [\mathbf{x}_T^T\mathbf{x}_T]^{1/2}. \tag{4.14}$$

Then for range rate, if we differentiate (4.14), and use (4.12) we get

$$\begin{aligned} \dot{r} &= \frac{\partial r}{\partial x_i}\,\dot{x}_i \\ &= \mathbf{x}_T^T\left[\mathbf{R}_{xq}\dot{\mathbf{q}} - \frac{\partial \mathbf{R}_{xu}}{\partial \theta}\,\mathbf{u}_0\dot{\theta}\right]\frac{1}{r}. \end{aligned} \tag{4.15}$$

For different types of observations, rotations to other coordinate systems shown in Figures 7 and 8 are convenient. The appropriate rotations can be deduced from Figures 7 and 8 by making one-at-a-time rotations in the same manner as was applied in deriving (2.24)–(2.30) and using the rules in (2.6) to write the rotation matrices. For coordinate **p**, with the 3-axis toward an inertially-fixed point of right ascension α and of declination δ (as would be appropriate for a photograph of a satellite against the stellar background), we have

$$\mathbf{p}_T = \mathbf{R}_{px}\mathbf{x}_T = \mathbf{R}_3\left(\frac{\pi}{2}\right)\mathbf{R}_2\left(\frac{\pi}{2} - \delta\right)\mathbf{R}_3(\alpha)\mathbf{x}_T$$

$$= \begin{bmatrix} -\sin\alpha, & \cos\alpha, & 0 \\ -\sin\delta\cos\alpha, & -\sin\delta\sin\alpha, & \cos\delta \\ \cos\delta\cos\alpha, & \cos\delta\sin\alpha, & \sin\delta \end{bmatrix}\mathbf{x}_T. \tag{4.16}$$

For coordinates **l** with the 3-axis toward the local zenith (ϕ, λ), we have

$$\mathbf{l}_T = \mathbf{R}_{lu}\mathbf{u}_T = \mathbf{R}_3\left(\frac{\pi}{2}\right)\mathbf{R}_2\left(\frac{\pi}{2} - \phi\right)\mathbf{R}_3(\lambda)\mathbf{u}_T. \tag{4.17}$$

If we are not concerned with orbital dynamics, the satellite can be treated merely as an elevated point, and the photograph itself can be used to obtain the orientation; a purely local coordinate system can be used rather than the externally referred $\mathbf{u}_T$ or $\mathbf{x}_T$. For a photograph of the satellite from the ground, the coordinate convention of terrestrial photogrammetry, shown in Figure 9, can be written as

$$\mathbf{b}_T = \mathbf{R}_3\left(\varkappa - \frac{\pi}{2}\right)\mathbf{R}_2\left(\frac{\pi}{2} - z\right)\mathbf{R}_1\left(-A - \frac{\pi}{2}\right)\mathbf{R}_2\left(-\frac{\pi}{2}\right)\mathbf{l}_T, \tag{4.18}$$

where $\varkappa$ is the swing, or roll, angle; z is the zenith distance; and A is the azimuth, measured clockwise from north. The coordinates of the image on a photograph taken by a camera of focal length f will then be fb_1/b_3 and fb_2/b_3. For a photograph taken from the satellite of the ground, the convention of aerial photogrammetry, shown in Figure 10, can be written as

$$\mathbf{b}_T = \mathbf{R}_3\left(\frac{\pi}{2} + s\right)\mathbf{R}_2\left(\frac{\pi}{2} - t\right)\mathbf{R}_1\left(\frac{\pi}{2} - A\right)\mathbf{R}_2\left(-\frac{\pi}{2}\right)\mathbf{m}_T, \tag{4.19}$$

where s, t, and A are known, respectively, as the swing, tilt, and azimuth. In the next section, in deriving differential relationships for use in observation equations, we shall refer back to Equations (4.14)–(4.19).

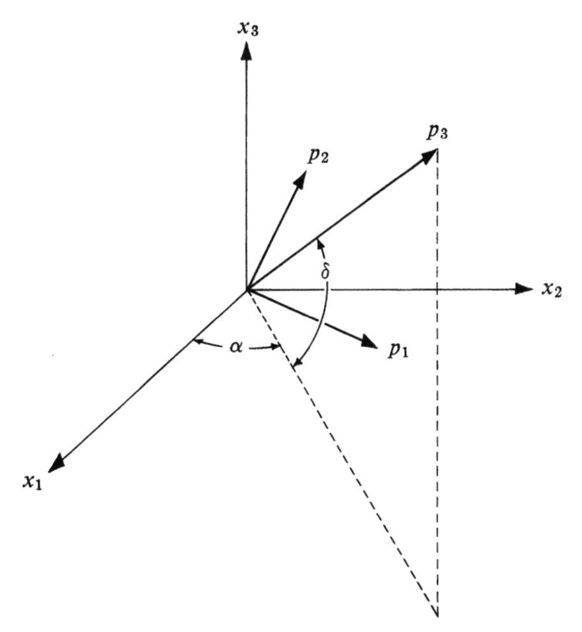

FIGURE 7. *Rotations from inertially-fixed to camera-axis coordinate systems.*

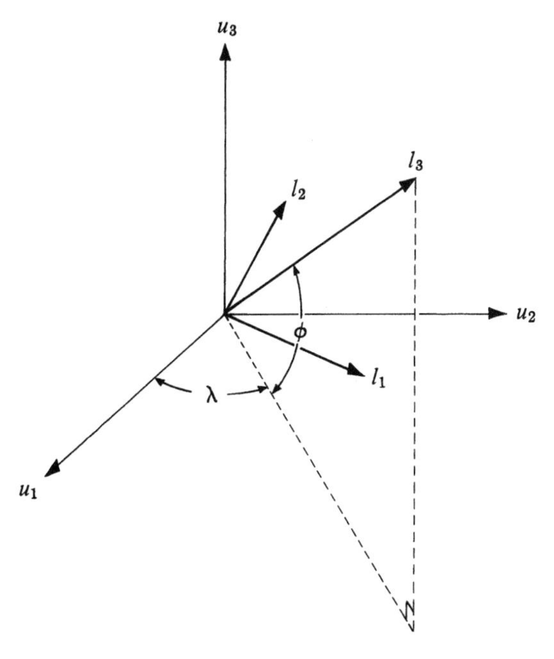

FIGURE 8. *Rotations from earth-fixed to local coordinate systems.*

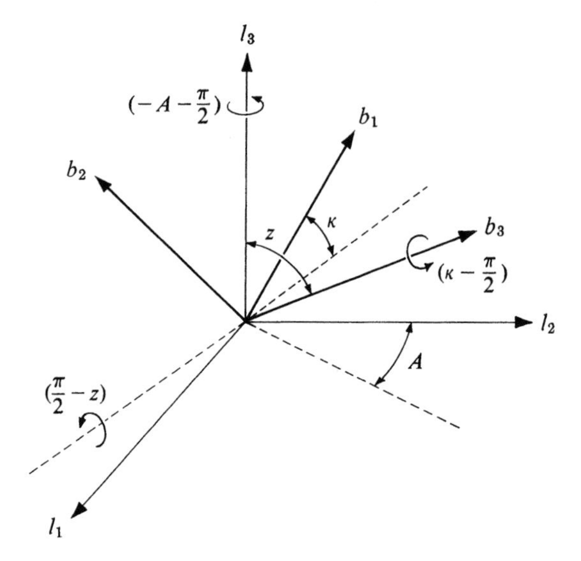

FIGURE 9. *Terrestrial photogrammetric coordinate rotations.*

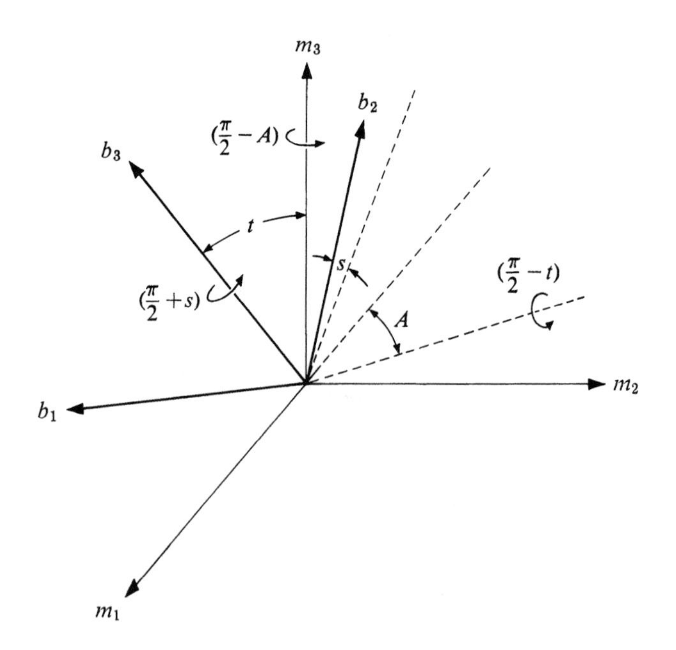

FIGURE 10. *Aerial photogrammetric coordinate rotations.*

4.3. Differential Relationships and Observation Equations

In order to apply the effects of known corrections to calculated directions, ranges, range-rates, and so on, that are to be compared to observations, or to determine these corrections from observations, we need to know the differential relationships of the directions, etc., to the parameters to be corrected.

In cases in which the satellite orbit dynamics are used, the intermediary coordinates used for all types of observations are the position and velocity in inertially referred rectangular coordinates $\mathbf{x}$ and $\dot{\mathbf{x}}$. Hence we require the partial derivatives of these coordinates with respect to the parameters determining the orbit: the six constants of integration; kM, J_2, and the smaller coefficients C_{lm}, S_{lm} of the gravitational field; and any parameters used to define a model of atmospheric drag or radiation pressure. In Chapter 3 all variations of the satellite position were expressed through the osculating Keplerian elements $\{a, e, i, M, \omega, \Omega\}$, hence we first require the partial derivatives $\mathbf{x}$ and $\dot{\mathbf{x}}$ with respect to these variables. For $\partial\mathbf{x}/\partial\mathbf{s}$, by differentiating (4.11) we have

$$d\mathbf{x} = \left[\frac{\partial \mathbf{R}_{xq}}{\partial(\Omega, i, \omega)}\mathbf{q} \;\vdots\; \mathbf{R}_{xq}\frac{\partial \mathbf{q}}{\partial(a, e, M)}\right]\begin{bmatrix} d\Omega \\ di \\ d\omega \\ da \\ de \\ dM \end{bmatrix}. \tag{4.20}$$

The derivatives of $\mathbf{R}_{xq}$ can be obtained by straightforward differentiation of (2.32), or by differentiating the appropriate components in turn of the matrix product, (2.31). For example, we have

$$\begin{aligned}\frac{\partial \mathbf{R}_{xq}}{\partial\Omega} &= \frac{\partial \mathbf{R}_3(-\Omega)}{\partial\Omega}\mathbf{R}_1(-i)\mathbf{R}_3(-\omega) \\ &= \begin{bmatrix} -\sin\Omega & -\cos\Omega & 0 \\ \cos\Omega & -\sin\Omega & 0 \\ 0 & 0 & 0 \end{bmatrix}\mathbf{R}_1(-i)\mathbf{R}_3(-\omega)\end{aligned} \tag{4.21}$$

$$= \begin{Bmatrix} -\sin\Omega\cos\omega - \cos\Omega\cos i\sin\omega, & \sin\Omega\sin\omega - \cos\Omega\cos i\cos\omega, & \cos\Omega\sin i \\ \cos\Omega\cos\omega - \sin\Omega\cos i\sin\omega, & -\cos\Omega\sin\omega - \sin\Omega\cos i\cos\omega, & \sin\Omega\sin i \\ 0 & 0 & 0 \end{Bmatrix}.$$

In getting derivatives of $\mathbf{q}$, care must be exercised that the dependence on the eccentricity through the eccentric anomaly E or the true anomaly f and range r are taken into account. On differentiating (3.13) and (3.14), and using $\partial E/\partial e$ from (3.19), we obtain

$$\frac{\partial \mathbf{q}}{\partial(a, e, M)} = \begin{bmatrix} \cos E - e, & -a\left(1 + \dfrac{\sin^2 E}{1 - e\cos E}\right), & -\dfrac{a\sin E}{1 - e\cos E} \\ \sqrt{1 - e^2}\sin E, & a\sqrt{1 - e^2}\sin E\left(\dfrac{\cos E}{1 - e\cos E} - \dfrac{e}{1 - e^2}\right), & \dfrac{a\sqrt{1 - e^2}\cos E}{1 - e\cos E} \\ 0 & 0 & 0 \end{bmatrix}$$

$$= \begin{bmatrix} q_1/a, & -a - \dfrac{q_2^2}{r(1 - e^2)} & -\dfrac{aq_2}{r\sqrt{1 - e^2}} \\ q_2/a, & \dfrac{q_1 q_2}{r(1 - e^2)}, & \dfrac{a}{r}\sqrt{1 - e^2}(q_1 + ae) \\ 0 & 0 & 0 \end{bmatrix}. \tag{4.22}$$

For the differential of the velocity $d\dot{\mathbf{x}}$, in terms of the osculating elements, differentiate (4.12):

$$d\dot{\mathbf{x}} = \left[\frac{\partial \mathbf{R}_{xq}}{\partial(\Omega, i, \omega)}\dot{\mathbf{q}} \;\middle|\; \mathbf{R}_{xq}\frac{\partial \dot{\mathbf{q}}}{\partial(a, e, M)}\right]\begin{bmatrix} d\Omega \\ di \\ d\omega \\ da \\ de \\ dM \end{bmatrix}, \tag{4.23}$$

of which the only new component is, from (3.24),

$$\frac{\partial \dot{\mathbf{q}}}{\partial(a, e, M)}$$

$$= \begin{Bmatrix} \dfrac{n\sin E}{2(1 - e\cos E)}, & \dfrac{na\sin E[e - 2\cos E + e\cos^2 E]}{(1 - e\cos E)^3}, & \dfrac{na(e - \cos E)}{(1 - e\cos E)^3} \\ \dfrac{-n\sqrt{1 - e^2}\cos E}{2(1 - e\cos E)}, & \dfrac{na[(\cos E - e)^2 - (1 - e\cos E)\sin^2 E]}{\sqrt{1 - e^2}\,(1 - e\cos E)^3}, & \dfrac{-na\sqrt{1 - e^2}\sin E}{(1 - e\cos E)^3} \\ 0 & 0 & 0 \end{Bmatrix}$$

$$= \begin{Bmatrix} -\dot{q}_1/2a, & \dot{q}_1\left(\dfrac{a}{r}\right)^2\left[2\left(\dfrac{q_1}{a}\right) + \dfrac{e}{1 - e^2}\left(\dfrac{q_2}{a}\right)^2\right], & -n\left(\dfrac{a}{r}\right)^3 q_1 \\ -\dot{q}_2/2a, & \dfrac{n}{\sqrt{1 - e^2}}\left(\dfrac{a}{r}\right)^2\left[\dfrac{q_1^2}{r} - \dfrac{1}{a(1 - e^2)}q_2^2\right], & -n\left(\dfrac{a}{r}\right)^3 q_2 \\ 0 & 0 & 0 \end{Bmatrix}. \tag{4.24}$$

As discussed in Chapter 3, the osculating elements $\{a, e, i, M, \omega, \Omega\}$ are functions of the constants of integration, the parameters of the gravitational field, kM and the coefficients C_{lm} and S_{lm}, and other parameters expressing the effects of the atmosphere, radiation pressure, and so on. To obtain the partial derivatives with respect to constants of integration that are "mean" Keplerian elements—namely, having no periodic perturbations, such as the parameters $\{a_0'', e_0'', i_0'', M_0'', \omega_0'', \Omega_0''\}$ of the final intermediary defined by the canonical transformation process of (3.105)–(3.113)—we may use the simplest possible expression of the orbit: the ellipse with constant a, e, i and node and perigee precessing due to the effect of the oblateness C_{20} as given by (3.74). Thus,

$$\frac{\partial(a,\ e,\ i)}{\partial(a_0'',\ e_0'',\ i_0'')} = \mathbf{I}, \quad \frac{\partial(a,\ e,\ i)}{\partial(M_0'',\ \omega_0'',\ \Omega_0'')} = \mathbf{O}, \quad \frac{\partial(M,\ \omega,\ \Omega)}{\partial(M_0'',\ \omega_0'',\ \Omega_0'')} = \mathbf{I},$$

and

$$\frac{\partial(M, \omega, \Omega)}{\partial(a_0'', e_0'', i_0'')}$$

$$= \begin{bmatrix} -\left[\dfrac{3n}{2a} + \dfrac{21\mu a_e^2 C_{20}(3\sin^2 i - 2)}{8na^6(1-e^2)^{3/2}}\right]\Delta t, & \dfrac{9\mu a_e^2(3\sin^2 i - 2)C_{20}e\,\Delta t}{4na^5(1-e^2)^{5/2}}, & \dfrac{9\mu a_e^2 C_{20}\sin 2i\,\Delta t}{4na^5(1-e^2)^{3/2}} \\ -\dfrac{21\mu a_e^2 C_{20}(1 - 5\cos^2 i)\,\Delta t}{8na^6(1-e^2)^2}, & \dfrac{3\mu a_e^2 C_{20}e(1 - 5\cos^2 i)\,\Delta t}{na^5(1-e^2)^3}, & \dfrac{15\mu a_e^2 C_{20}\sin 2i\,\Delta t}{4na^5(1-e^2)^2} \\ -\dfrac{21\mu a_e^2 C_{20}\cos i\,\Delta t}{4na^6(1-e^2)^2}, & \dfrac{6\mu a_e^2 C_{20}e\cos i\,\Delta t}{na^5(1-e^2)^3}, & -\dfrac{3\mu a_e^2 C_{20}\sin i\,\Delta t}{2na^5(1-e^2)^2} \end{bmatrix}. \tag{4.25}$$

Equation (4.25) is obtained by differentiating (3.74), and the $\partial n/\partial a_0''$ of $-3n/2a$ by differentiating Kepler's third law, (3.20). Δt in (4.25) is the time difference between the instant of observation and the epoch to which the elements a_0'', e_0'', i_0'' refer. If this time difference Δt is such that the elements in (4.25) other than $-3n/2a$ are small compared to unity—less than 10 days or so—these elements may be neglected.

For the partial derivatives with respect to constants of integration which are osculating Keplerian elements at epoch a_0, e_0, i_0, M_0, ω_0, Ω_0, an additional Jacobian $\partial(a_0'', e_0'', i_0'', M_0'', \omega_0'', \Omega_0'')/\partial(a_0, e_0, i_0, M_0, \omega_0, \Omega_0)$ must be applied to the right of (4.25). The off-diagonal elements of this array will be derivatives of periodic perturbations, and hence will contain C_{20} as a factor. Therefore they will be significant only where their multiplier (4.25) may be large compared to unity: namely, $\partial M/\partial a_0''$. So for the partial derivative of the mean anomaly M with respect to any osculating element s_{i0} at epoch there must be added

$$-\frac{3n}{2a}\Delta t\,\frac{\partial\,\Delta a_{20}}{\partial s_{i0}},$$

where, from (3.76), assuming only $\dot{M}$ significant in the denominator,

$$\Delta a_{20} = \frac{2a_e^2 C_{20}}{a} \sum_{pq \neq 10} F_{20p} G_{2pq} \cos \{(2 - 2p)\omega_0 + (2 - 2p + q)M_0\}.$$

For the partial derivatives of the osculating elements with respect to the coefficients of the gravitational field, (3.76) would be used. For example, substituting from (3.71) for S_{lmpq}, we obtain

$$\frac{\partial a}{\partial C_{lm}} = \mu a_e^l \sum_{p,q} \frac{2F_{lmp}G_{lpq}(l - 2p + q)}{na^{l+2}[(l - 2p)\dot{\omega} + (l - 2p + q)\dot{M} + m(\dot{\Omega} - \dot{\theta})]}$$
$$\times \begin{bmatrix} \cos \\ \sin \end{bmatrix}_{(l-m)\text{ odd}}^{(l-m)\text{ even}} [(l - 2p)\omega + (l - 2p + q)M + m(\Omega - \theta)]. \quad (4.26)$$

Because the magnitude of the effects of the coefficients C_{lm}, S_{lm} on the elements and the variations in a, e, i are both small, we can write the partial derivatives such as (4.26) as a sum of sinusoidal functions with constant coefficients and arguments having a constant rate of change with respect to time:

$$\frac{\partial s_i}{\partial(C_{lm} \text{ or } S_{lm})} = \sum_{p,q} K_{ilmpq}(a, e, i) \begin{bmatrix} \cos \\ \text{or} \\ \sin \end{bmatrix} [(l - 2p)(\omega_0 + \dot{\omega}\,\Delta t)$$
$$+ (l - 2p + q)(M_0 + \dot{M}\,\Delta t) + m(\Omega_0 + \dot{\Omega}\,\Delta t - \theta_0 - \dot{\theta}\,\Delta t)], \quad (4.27)$$

where, for example, for $s_i = a$,

$$K_{ilmpq} = \mu a_e^l \frac{2F_{lmp}G_{lpq}(l - 2p + q)}{na^{l+2}\{(l - 2p)\dot{\omega} + (l - 2p + q)\dot{M} + m(\dot{\Omega} - \dot{\theta})\}}.$$

Equation (3.76), on which (4.27) is based, was derived on the assumption that the perturbations by the gravity field are forced oscillations superimposed on a secularly moving ellipse. Hence (4.27) is valid in cases in which the constants of integration do not include perturbations, that is, "mean" elements such as the parameters of the final intermediary defined by the canonical transformation process of (3.105)–(3.113).

If, however, the manner of calculating the orbit uses constants of integration that include the effects of perturbations, as would be the case in a numerical integration starting from osculating elements or position and velocity at epoch, then the effect of the perturbation at epoch must be subtracted out; that is, (4.27) must be replaced by

$$\frac{\partial s_i}{\partial(C_{lm} \text{ or } S_{lm})} = \sum_{p,q} \Bigg\{ K_{ilmpq}(a,\ e,\ i) \begin{bmatrix} \cos \\ \text{or} \\ \sin \end{bmatrix} [(l-2p)(\omega_0 + \omega\,\Delta t)$$
$$+ (l-2p+q)(M_0 + \dot{M}\,\Delta t) + m(\Omega_0 + \dot{\Omega}\,\Delta t - \theta_0 - \dot{\theta}\,\Delta t)]$$
$$- \sum \frac{\partial s_i}{\partial s_{j0}} K_{jlmpq}(a,\ e,\ i) \begin{bmatrix} \cos \\ \text{or} \\ \sin \end{bmatrix} [(l-2p)\omega_0$$
$$+ (l-2p+q)M_0 + m(\Omega_0 - \theta_0)] \Bigg\}. \tag{4.28}$$

Combining (4.20), (4.23), (4.25), and (4.27) or (4.28), we can write

$$d\begin{bmatrix} \mathbf{x} \\ \dot{\mathbf{x}} \end{bmatrix} = \frac{\partial\{\mathbf{x},\ \dot{\mathbf{x}}\}}{\partial\{a,\ e,\ i,\ M,\ \omega,\ \Omega\}} \Bigg[\frac{\partial\{a,\ e,\ i,\ M,\ \omega,\ \Omega\}}{\partial\{a_0'',\ e_0'',\ i_0'',\ M_0'',\ \omega_0'',\ \Omega_0''\}}\, d \begin{bmatrix} a_0'' \\ e_0'' \\ i_0'' \\ M_0'' \\ \omega_0'' \\ \Omega_0'' \end{bmatrix}$$
$$+ \sum_{l,m} \frac{\partial\{a,\ e,\ i,\ M,\ \omega,\ \Omega\}}{\partial\{C_{lm},\ S_{lm}\}} \begin{pmatrix} dC_{lm} \\ dS_{lm} \end{pmatrix}$$
$$+ \sum \frac{\partial\{a,\ e,\ i,\ M,\ \omega,\ \Omega\}}{\partial\{\text{other parameters}\}}\, d\{\text{other parameters}\} \Bigg]. \tag{4.29}$$

For each type of observation, we require the appropriate modification of one of the range, range rate, or position vectors (4.14)–(4.19) to obtain the calculated equivalent of the quantities observed. In general, for any observation designated by subscript i,

$$\text{Obs}_i \neq \text{Calc}_i. \tag{4.30}$$

The procedure of differential correction is to determine corrections to observations $d\,\text{Obs}$ and corrections to parameters $d\,\text{Par}$ designated by subscript j such that

$$\text{Obs}_i + d\,\text{Obs}_i = \text{Calc}_i + \sum_j \frac{\partial\,\text{Calc}_i}{\partial\,\text{Par}_j}\, d\,\text{Par}_j. \tag{4.31}$$

On abbreviating the expressions in (4.31) and writing it in its more usual form of residuals "O–C", (4.31) becomes

$$(O - C)_i = \sum_j \frac{\partial C_i}{\partial P_j} dP_j - dO_i. \tag{4.32}$$

Some of the parameters P_j are, as we have discussed, those that affect the orbit. The effect of these parameters on the calculated value C_i for a particular observation can be represented entirely by the position $\mathbf{x}$ or velocity $\dot{\mathbf{x}}$ at the instant the signal that is measured leaves the satellite; that is, we can write

$$\frac{\partial C_i}{\partial P_j} = \frac{\partial C_i}{\partial x_k} \cdot \frac{\partial x_k}{\partial P_j} + \frac{\partial C_i}{\partial \dot{x}_k} \cdot \frac{\partial \dot{x}_k}{\partial P_j}, \tag{4.33}$$

where the k subscript is summed from 1 to 3 and the derivatives $\partial x_k/\partial P_j$ and $\partial \dot{x}_k/\partial P_j$ are those given by (4.29). With one exception, these derivatives are about all that can readily be considered as common to all types of observations. This one exception is the effect of a timing correction or error $\epsilon(t)$. Since the effect of such an error is to cause the calculated position of the satellite to be too far, or not far enough, along the orbit, it can be calculated as the effect of a variation in the mean anomaly dM multiplied by the rate of change of M with respect to time; that is, the mean motion n as calculated by Kepler's law (3.20). Thus,

$$\frac{\partial C_i}{\partial \epsilon(t_i)} = \left[\frac{\partial C_i}{\partial x_k} \cdot \frac{\partial x_k}{\partial M} + \frac{\partial C_i}{\partial \dot{x}_k} \cdot \frac{\partial \dot{x}_k}{\partial M}\right] n, \tag{4.34}$$

where $\partial x_k/\partial M$ or $\partial \dot{x}_k/\partial M$ are calculated by (4.20) or (4.23), respectively. $\partial C_i/\partial \epsilon(t_i)$ as calculated by (4.34) is also used to make the correction for the travel time of the signal, or "planetary aberration" effect. On calculating the range r by (4.14), this correction will be

$$dC_{i,a} = -\frac{\partial C_i}{\partial \epsilon(t_i)} \cdot \frac{r}{c}, \tag{4.35}$$

where c is the velocity of light.

In cases where simultaneous observations are made of the satellite and the orbit is not used, the position x_k or velocity $\dot{x}_k$ coordinates themselves become parameters; that is, $\partial x_k/\partial P_i$ and $\partial \dot{x}_k/\partial P_j$ in (4.33) are identity matrices.

It remains therefore to consider separately for each type of observation, proceeding from left to right in (4.32):

1. The nature of the observation O_i as it enters the mathematical formulation of the problem: in what manner, if at all, the pure physically measured quantities are modified or transformed or combined before actually being used in the calculations.

2. The computation of the quantity C_i to be compared with the observation O_i. In the computation of C_i, various corrections may be applied that, from (4.32), obviously could just as well be applied with reversed sign to O_i, and vice versa. In this category are usually corrections that are caused by environmental effects such as refraction, and so forth. Since one series of observations O_i will normally be used in the determination of several alternative sets of model parameters P_j, it is computationally most convenient (a) to apply as corrections to the observations O_i those corrections that do not vary by any likely change in parameters P_j, and (b) to apply as corrections to the calculated quantities C_i those effects that may vary perceptibly with the likely change in the parameters P_j.

3. The selection of a mathematical model by selection of a set of parameters P_j to be corrected, and the calculation of the partial derivatives $\partial C_i/\partial P_j$.

4. The "correction to observation" dO_i obviously must, from (4.32), account for all of the discrepancy $(O - C)_i$ that cannot be accounted for by the corrections to parameters dP_j. Hence it must absorb all of the discrepancy that is caused by the incompleteness of the mathematical model represented by parameters P_j. Thus dO_i can be very much affected by things that are not at all errors in observation in the usual sense. In the case of interest here, close satellite orbits, this situation will very likely exist, because of the inadequacy of our model of drag effects on the orbit. Given the state of our knowledge, it is inevitable that such situations will occur. Since such errors arise from the environment, over which we have much less control than over instrumentation, they are much more likely to have a correlated character: that is, the closer together errors are in time, the more similar they usually will be in sign and magnitude. The statistical implications of such correlation we shall take up in the next chapter; in this chapter in discussing the differential corrections of observations resulting from different effects, we still want to consider the order of magnitude and degree of correlation of these effects in order to decide where and how certain corrections should be applied as well as to decide the appropriate statistical treatment.

4.4. Observation Equations: Directional

The principal directional method of tracking satellites is to photograph the satellite against the background of stars. The actual measurements made are the rectangular coordinates of the images of the satellite and the stars on the photograph, and the instants of time at which the film was exposed, as determined from standard time signals. The geometry of the camera is shown in Figure 11. The coordinates (x_o, y_o) measured on the photograph can thus be seen to be the projection at the focal distance f from the focus of the camera of the rectangular coordinates of the satellite referred to a system with the 3-axis coinciding with the camera axis. If the axis of the camera points toward right ascension and declination (α, δ), as determined from the catalogued coordinates of the stars on the photograph, then the computed coordinates (x_c, y_c) can be obtained from (4.16). Therefore

$$C_i = \begin{bmatrix} x_c \\ y_c \end{bmatrix} = \begin{bmatrix} f/p_3 & 0 & 0 \\ 0 & f/p_3 & 0 \end{bmatrix} \mathbf{R}_{px}\mathbf{x}_T, \tag{4.36}$$

in which the coordinate p_3 coincides with the range r. The coordinates (x_c, y_c) are often called standard coordinates, and symbolized by (ξ, η).

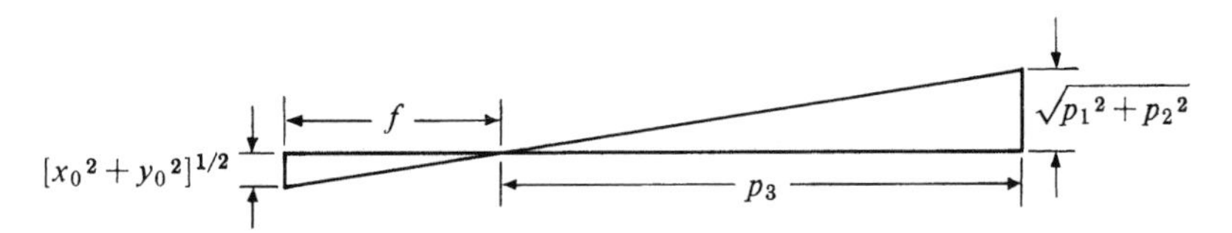

FIGURE 11. *Camera geometry.*

Quantities that may be applied as known corrections are:

1. The changes in right ascension and declination (α, δ) of the stars caused by precession and nutation from the time to which the star catalogue refers to the time to which the calculation in (4.36) is referred. This shift is large, but precisely known, as described later in Section 4.6. The time to which the calculations are referred is the epoch of the constants of integration of the orbit, which is normally central in time to the set of observations analyzed, for the reason that we want to keep the dynamical calculations as simple as possible by using an inertial frame in which the principal perturbation—the earth's oblateness—will be symmetric about the equator. This time is peculiar to the orbital model, rather than to the observations, and hence the shift is considered as a correction to the calculated quantities C_i, or $\{x_c, y_c\}$, rather than the observed quantities O_i, or $\{x_o, y_o\}$. A major part of the shift—from the catalogue epoch to the instant of observation—is sometimes, however, applied to the observed quantities O_i, but since it is a known correction the saving in thus reducing the correction to C_i is of little significance.

2. Time signal corrections, as issued some time subsequent to the observation by standard time services. See Section 4.6.

Quantities that must be considered as unknown errors include, roughly in sequence of magnitude:

1. Errors in orbital parameters, as expressed through the right side of (4.29).

2. Error in coordinates of the camera, either relative to the earth's center of mass or to another station on the earth's surface (as would be of interest in simultaneous observations of the satellite not using the orbit).

The differential expression dC_i combining these two categories of error can be obtained by substituting for $\mathbf{x}_T$ in (4.36) from (4.11) and differentiating. Thus

$$dC_i = \begin{bmatrix} f/p_3 & 0 & 0 \\ 0 & f/p_3 & 0 \end{bmatrix} \mathbf{R}_{px}[d\mathbf{x} - \mathbf{R}_3(-\theta)d\mathbf{u}_0] \tag{4.37}$$

In (4.37) $d\mathbf{u}_0$ is the correction to station position; the satellite position shift $d\mathbf{x}$ would be expressed by (4.29) when the orbit was used, or would be considered a set of 3 corrections to parameters itself in a system of simultaneous observations; and third-column terms $-fp_1/p_3^2$, $-fp_2/p_3^2$ have been set zero, being negligible.

3. Error in timing, resulting either from anomalies in travel of the time signal or to imperfect synchronization of a camera shutter or satellite light flash with the time reference. Such errors should not be more than $\pm 0.001^s$ or $\pm 0.002^s$; if significant, they can be considered as equivalent to errors in the direction of motion of the satellite image, with partial derivatives as calculated by (4.34).

4. Error in the measured coordinates of the satellite on the photo plate, resulting either from anomalies in atmospheric refraction ("shimmer") or to irregularities in motion by a tracking telescope, or to the plate measurement itself. Regardless of the source of the error, the significance of the error is a purely differential one. That is, the satellite image is affected differently from the star images. Since the atmospheric shimmer and tracking irregularities are rapidly time varying, these sources of error will be more important when the satellite image is impressed at a time different from the star images. These sources of error will be most important when it is imposed in a very short time, that is, when the light source is a flash. Errors of this sort appear to have a magnitude of $\pm 2''$ or $3''$, the shimmer error being to some extent inversely correlated with camera aperture. On the other hand, plate measurement error will be most significant when the character of the satellite image differs the most from the star images, as, for example, in a sidereally mounted camera where the star images are points and the satellite images are breaks in a trail. Errors of this sort vary from ± 2 to 5 microns, or from $\pm 0.5''$ to $2.0''$. If there are several images of each class—satellite and star—on the same plate, then it may be desirable to include as additional

unknowns in the reduction of plate measurements differential biases Δx, Δy between the classes of images.

5. Error in the catalogue position of stars may be $\pm 0.5''$, or in some cases it may be desirable to use multiple images of an uncatalogued star, in which case the star position constitutes two additional unknowns.

6. Exterior orientation of the camera. In order to establish the orientation of the camera axes with respect to an external system, three angles are required. If the observations are made with an equatorially mounted camera (designed to keep the camera axes fixed with respect to inertial space), then they are appropriately defined in terms of the $\mathbf{p}_T$ vector, (4.16). In this case, the orientation angles are the right ascension α and declination δ of the camera axis, and the discrepancy $d\varkappa$ in roll, or swing, about the p_3 axis between the assumed direction of the p_1 axis and the actual direction of north from (α, δ). If the observations are made with a ground-fixed camera (designed to keep the camera axes fixed with respect to the earth), then they are appropriately defined in terms of the $\mathbf{b}_T$ vector, (4.18). In this case, the orientation angles are the swing $\varkappa$, the zenith distance z, and the azimuth A, as shown in Figure 9.

In either case, to obtain the orientation angles for a particular photo plate by comparison of calculated and measured plate coordinates of several stars, each star of right ascension and declination (α, δ) can be assumed to be on the unit sphere. Thus

$$\mathbf{x}_s = \begin{cases} \cos\delta\cos\alpha, \\ \cos\delta\sin\alpha, \\ \sin\delta, \end{cases} \tag{4.38}$$

and the computed plate coordinates (x_c, y_c) obtained by appropriate modification of (4.36) are

$$\begin{bmatrix} x_c \\ y_c \end{bmatrix} = \begin{bmatrix} f/p_3 & 0 & 0 \\ 0 & f/p_3 & 0 \end{bmatrix} \mathbf{R}_3(d\varkappa)\mathbf{R}_{px}\mathbf{x}_s, \tag{4.39}$$

for equatorially mounted cameras, and

$$\begin{bmatrix} x_c \\ y_c \end{bmatrix} = \begin{bmatrix} f/p_3 & 0 & 0 \\ 0 & f/p_3 & 0 \end{bmatrix} \mathbf{R}_{bl}\mathbf{R}_{lu}\mathbf{R}_{ux}\mathbf{x}_s, \tag{4.40}$$

for ground-fixed cameras, where $\mathbf{R}_{bl}$, $\mathbf{R}_{lu}$, and $\mathbf{R}_{ux}$ are obtained from (4.18), (4.17), and (2.27), respectively.

7. Interior orientation of the camera, or centering and scale. The center of the photographic plate as established by fiducial marks used as a reference

in comparator measurements may not coincide with the axis of the camera. Hence it is necessary to determine the coordinates x_p, y_p of the camera axis, called the principal point, relative to the plate coordinates. Also the focal length f must be known to obtain the correct scale relationship between location of the photo image (x_o, y_o) and the exterior location of the object, $\mathbf{p}_T$ or $\mathbf{b}_T$. Both the centering (x_p, y_p) and scale factor f are relatively constant and hence determined only at occasional calibrations.

8. Camera distortion. In a properly designed camera, distortion by the lens should amount to less than 10 microns, and be purely radial from the principal point. Hence it can be expressed by a polynomial in odd powers of the radial coordinate r (even powers being eliminated by symmetry and continuity considerations). Thus

$$\delta_v = k_0 r + k_1 r^3 + k_2 r^5 + k_3 r^7 + \cdots . \tag{4.41}$$

The leading coefficient k_0 is indistinguishable in effect from the scale factor f. The distortion is constant, so the other coefficients k_1, k_2, k_3, $\cdots$ can be determined at occasional calibrations with the interior orientation and scale parameters. Then to correct measured coordinates (x_o, y_o),

$$\begin{bmatrix} \Delta x_o \\ \Delta y_o \end{bmatrix} = \begin{bmatrix} x_o \\ y_o \end{bmatrix} \frac{\delta r}{r} . \tag{4.42}$$

9. Differential atmospheric refraction. Near the horizon, the vertical component of atmospheric refraction may vary significantly between the satellite image and the star images. In this case, the atmospheric refraction may be represented by a power series in the tangent of the zenith distance, and the coefficients of the power series may be considered as additional unknown parameters. Being a function of zenith distance, refraction parameters require a locally referred coordinate system.

In observations of a satellite, the errors (3) through (9) are all peculiar to a particular camera or to a particular pass. These errors enter into an equation, (4.37), with corrections to parameters external to the particular observation only through the independent quantities $\{x_o, y_o\}$. The two rotations $\{\alpha, \delta\}$ used in the matrix $\mathbf{R}_{px}$ that are also determined from the photograph are coupled with $\{x_o, y_o\}$. Because tracking cameras have narrow fields of view, in publication of camera data it is in fact customary to consider the camera satellite line α, δ as the camera axis, in which case the $\{x_o, y_o\}$ are zero. It would be manifestly undesirable to complicate an adjustment involving shifts to the satellite position $d\mathbf{x}$ and to the station position $d\mathbf{u}_0$ by combining it with corrections for camera orientation, refraction, and so forth. Hence distortion and internal orientation parameters are generally determined in occasional calibrations; external orientation, differential bias, and refraction are determined in a preliminary adjustment for each photographic plate, and the results expressed as a final $\{x_o, y_o\}$ and $\{\alpha, \delta\}$ for use in (4.36) and (4.37). See Brown (1964) for full details.

There are two types of directional observations of lesser accuracy than cameras:

First there are theodolites that measure zenith distance and azimuth. The computed value is most conveniently obtained from the $\mathbf{l}_T$ vector components, where $\mathbf{l}_T$ is calculated by (4.17):

$$C_i = \begin{bmatrix} z_c \\ A_c \end{bmatrix} = \begin{bmatrix} \tan^{-1} \dfrac{\sqrt{l_1^2 + l_2^2}}{l_3} \\[2ex] \tan^{-1} \dfrac{l_1}{l_2} \end{bmatrix}, \tag{4.43}$$

whence

$$(O - C)_i = \begin{bmatrix} 1/r & 0 & 0 \\ 0 & \csc z/r & 0 \end{bmatrix} \mathbf{R}_3\left(-\frac{\pi}{2}\right) \mathbf{R}_1(-z) \mathbf{R}_3(-A) \mathbf{R}_{lu} \times [R_3(\theta)\, d\mathbf{x} - d\mathbf{u}_0] - dO_i. \tag{4.44}$$

Second, there are interferometers that determine the direction cosine with respect to an axis established by two radio antennas that measure the phase difference of a radio signal received from the satellite. The calculated direction cosine is

$$C_i = \left[\frac{\sin A}{r},\ \frac{\cos A}{r},\ 0\right] \mathbf{l}_T, \tag{4.45}$$

where A is the azimuth of the axis, calculated from l_1, l_2 as in (4.39), and $\mathbf{l}_T$ is calculated by (4.17). The observation equation is

$$(O - C)_i = \mathbf{N}' \mathbf{R}_{lu} [\mathbf{R}_3(\theta)\, d\mathbf{x} - d\mathbf{u}_0] - dO_i, \tag{4.46}$$

in which

$$\mathbf{N}' = \left[\frac{\sin A}{r} - \frac{l_1^2 \sin A + l_1 l_2 \cos A}{r^3},\ \frac{\cos A}{r} - \frac{l_1 l_2 \sin A + l_2^2 \cos A}{r^3},\ - \frac{l_3(l_1 \sin A + l_2 \cos A)}{r^3}\right]. \tag{4.47}$$

4.5. Observation Equations: Range Rate and Range

The range rate of a satellite is determined by the Doppler technique, in which the shift in frequency of a radio signal caused by motion of the source is measured. The actual measurements made are the counts of Doppler cycles over a certain standard interval, such as one second. Division of the count by the length of the interval thus yields the mean Doppler frequency for the interval. Given the reference frequency f_i and the velocity of light c, the Doppler frequency Δf_i can be related to the range rate $\dot{r}$ by

$$\Delta f_i = \frac{f_i}{c}\dot{r} + \frac{\alpha}{f_i}. \tag{4.48}$$

The second term in (4.48) expresses ionospheric refraction effect; it is inversely proportional to f_i because the effect of an ionized medium on the velocity of a radio signal is inversely proportional to f_i^2. The ion density at any instant is sufficiently unsure that the parameter α must be considered an unknown. Hence two frequencies f_i, $i = 1, 2$, are commonly employed and two equations such as (4.48) solved simultaneously to obtain the observed range rate $\dot{r}_o$:

$$O_i = \dot{r}_o = c\frac{f_1\,\Delta f_1 - f_2\,\Delta f_2}{f_1^2 - f_2^2}. \tag{4.49}$$

The calculated range rate C_i is given by (4.15). In addition to time signal error, a quantity which is applied as a known correction to the observed range rate $\dot{r}_o$ is tropospheric refraction. The effect of this refraction on a range is the integral of the refractive index μ over the ray path; hence the effect on the range rate will be the time rate of change of this integral:

$$\begin{aligned}\delta\dot{r} &= -\frac{d}{dt}\int_0^r \mu\,ds = -\int_0^r \frac{d\mu}{dt}\,ds \\ &= -\int_0^r \frac{d\mu}{dh}\cdot\frac{dh}{dz}\cdot\frac{dz}{dt}\,ds.\end{aligned} \tag{4.50}$$

Here ds is an increment along the ray path (negligibly different for this purpose from the station-satellite line); z is zenith distance, calculated from $\mathbf{l}_T$ as in (4.39); and h is height above the earth, calculated by (see Figure 12)

$$h = \sqrt{R^2 + s^2 + 2Rs\cos z} - R. \tag{4.51}$$

The time differentiation in (4.50) can be moved inside the integral because all change in the upper limit to s takes place at the satellite, where there is no tropospheric refraction. The refractive effect $(\mu - 1)$ is generally expressed as an exponential function of h. The rate of change dz/dt is computed from the orbital motion. All these complications make it more practicable to calculate $\delta\dot{r}$ by numerical integration. $\delta\dot{r}$ is less than 10 cm/sec for satellites

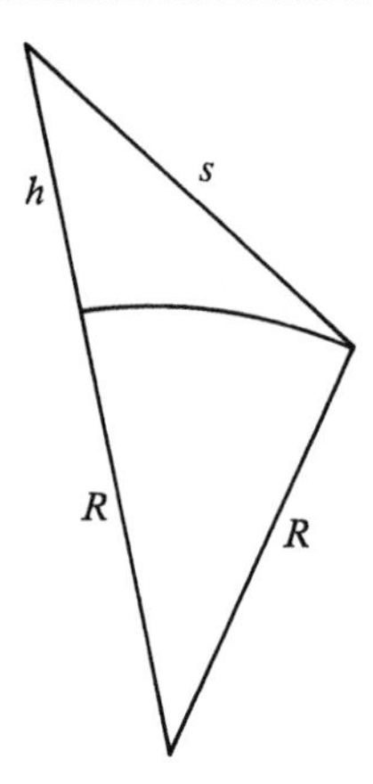

FIGURE 12. *Altitude–distance relationship.*

more than about 15° above the horizon, but increases sharply below this altitude. Since the uncertainty in $(\mu - 1)$ is about 10 per cent of its magnitude, it is thus desirable to reject observations less than about 15° above the horizon.

For the differential correction of the computed range rate caused by satellite velocity and station position, differentiate (4.15):

$$dC_i = d\dot{r}_i = d\left(\frac{\mathbf{x}_T \cdot \dot{\mathbf{x}}_T}{r}\right)$$

$$= \frac{1}{r}\,\mathbf{x}_T \cdot \left[d\dot{\mathbf{x}} - \frac{\partial \mathbf{R}_3(-\theta)}{\partial \theta}\,d\mathbf{u}_0\dot{\theta}\right]$$

$$+ \left[\frac{d\mathbf{x} - \mathbf{R}_3(-\theta)\,d\mathbf{u}_0}{r} - \frac{\mathbf{x}_T \cdot (d\mathbf{x} - \mathbf{R}_3(-\theta)\,d\mathbf{u}_0)}{r^3}\,\mathbf{x}_T\right] \cdot \dot{\mathbf{x}}_T. \quad (4.52)$$

Other quantities that must be considered as unknown errors, but are normally adjusted at a preliminary stage include:

1. Error in the reference frequency generated in the satellite constant throughout the pass;

2. Sometimes, error in the reference frequency that drifts throughout the pass;

3. Higher-order ionospheric refraction effect, not accounted for by the parameter α of (4.48);

4. Variations of the tropospheric refraction from that calculated by the model of $d\mu/dh$ used in (4.50);

5. "Noisy" or irregular data points, caused by failure to "lock" onto the signal and other instrumental effects.

The correction of these observational errors is usually combined with the process of aggregation. A typical pass of Doppler data may include as many as 400 data points. However, the significant geodetic information can generally be expressed by less than six numbers per pass—probably on the order of three. Since the range rate varies sharply within a pass, it is not practicable to determine arbitrary parameters from the range-rate itself, but rather to determine the arbitrary parameters from the residuals of the observed range rates $\dot{r}_o$ with respect to calculated range rates $\dot{r}_c$ based on a nominal reference orbit. For example, for n points and $m + 1$ parameters, the least-squares condition

$$\sum_{i=1}^{n}\left[(\dot{r}_{oi} - \dot{r}_{ci}) - \sum_{j=0}^{m} a_j(t_i - t_0)^j\right]^2 = \min, \tag{4.53}$$

where t_0, the time of the midpoint of the pass, can be used to determine the polynomial coefficients a_j. Then at a few selected times t_k the aggregated range rate $\dot{r}_a$ will be

$$\dot{r}_a = \dot{r}_c(t_k) + \sum_{j=0}^{m} a_j(t_k - t_0)^j. \tag{4.54}$$

The polynomial fitting will also eliminate some error by smoothing. The aggregated range rates can then be used as the observations O_i in the observation equation (4.32), in which the computed partial derivatives $\partial C_i/\partial P_j$ are calculated by a combination of (4.29) and (4.52). It is not necessary that the pass be finally expressed in terms of aggregated range rates $\dot{r}_a$; other parameters could be used, such as along- and across-track apparent station position errors and the mean frequency discrepancy. See Guier (1963a,b) and Hopfield (1963) for details on the reduction of Doppler tracking data.

For range measurements, the actual measurements are either time delays between transmitted and received radar pulses or the phase shifts in the modulation of a received signal with respect to a coherent transmitted signal. Range measurements will either be at frequencies in excess of 10^3 Mc/sec or else overcome ionospheric refraction by use of dual frequencies similar to (4.48)–(4.49). They will also require tropospheric refraction corrections and aggregation similar to (4.47) and (4.50)–(4.51). The calculated range is obtained from (4.14), whence the correction thereto becomes

$$dr = \frac{1}{r}\,\mathbf{x}_T \cdot [d\mathbf{x} - \mathbf{R}_3(-\theta)\, d\mathbf{u}_0]. \tag{4.55}$$

4.6. Time and Precise Definition of Coordinates

Time is measured by counting some type of repeated phenomenon. The ideal time to use as a reference for observations of satellite orbits would be a "gravitational" time: that is, one which exactly coincides with the time used as the independent variable in the dynamical developments of Chapter 3. If all orbital motions are consistent with the laws on which the dynamical developments are based—as they have so far been observed to be, taking into account relativistic effects where necessary—then the ideal repeated phenomenon to count for a time standard would be the most accurately known orbital period. This most accurately known period is that of the moon's revolution around the earth. The time measured by the moon's motion is known as Ephemeris Time (ET). (Though best measured by the moon's motion, ET is formally defined by the period of the earth's motion around the sun for the year 1900.0) Ephemeris Time is not immediately available for timing artificial satellite observations, however. Instead, the time services provide Atomic Time A1, measured by the resonant frequency of oscillation of the cesium atom. A1 time has thus far been found to be indistinguishable from ET and serves as an entirely satisfactory substitute for geodetic satellite purposes.

The equations written thus far assume that the coordinate system $\mathbf{x}$ defined by the x_1 axis toward the vernal equinox and the x_3 axis along the earth's rotation axis is fixed with respect to inertial space. In fact, the directions of the rotation axis and the vernal equinox are continually changing, as a result of the precession and nutation of the earth due to the attraction of the moon for the earth's bulge. At any time t the equator-equinox referred coordinates $\mathbf{x}$ (called true coordinates) of a point fixed with respect to the earth's center and an inertial frame can be represented by a differential rotation $\mathbf{R}_{x\bar{x}}$ applied to coordinates $\bar{\mathbf{x}}$ (called mean coordinates) that have purely secular change (namely, a polynomial of t) only. $\mathbf{R}_{x\bar{x}}$ being a differential rotation, we can take cosines to be 1, sines to be equal to the angles, and products of sines to be 0. Under these conditions, a rotation can be expressed as a product of rotations about each of the three coordinate axes in any order. It is convention to apply a counterclockwise rotation about the 2-axis: $\Delta\nu$ (called the nutation in declination) and clockwise rotations about the 1-axis and the 3-axis: $-\Delta\epsilon$ (called the nutation in obliquity) and $-\Delta\mu$ (called the nutation in right ascension), respectively. See Figure 13. Then

$$\mathbf{R}_{x\bar{x}} = \mathbf{R}_1(-\Delta\epsilon)\mathbf{R}_2(\Delta\nu)\mathbf{R}_3(-\Delta\mu)$$
$$= \begin{bmatrix} 1 & -\Delta\mu & -\Delta\nu \\ \Delta\mu & 1 & -\Delta\epsilon \\ \Delta\nu & \Delta\epsilon & 1 \end{bmatrix}. \tag{4.56}$$

The mean coordinates $\bar{\mathbf{x}}(t)$ are conventionally expressed in terms of the mean coordinates at an epoch t_o, $\bar{\mathbf{x}}(t_o)$ through a clockwise rotation about the 3-axis, $-(\varkappa + \omega)$, called precession in right ascension, and a counter-clockwise rotation about the 2-axis, ν, called precession in declination. The precession in right ascension is split into two equal parts, half-$\varkappa$ along the mean equator of t_0 and half-ω along the mean equator at t. See Figure 14. Then

$$\mathbf{R}_{\bar{x}\bar{x}_0} = \mathbf{R}_3(-\omega)\mathbf{R}_2(\nu)\mathbf{R}_3(-\varkappa). \tag{4.57}$$

We have expressed the effects of precession and nutation as being applied to rectangular coordinates referred to equator and equinox. In practice

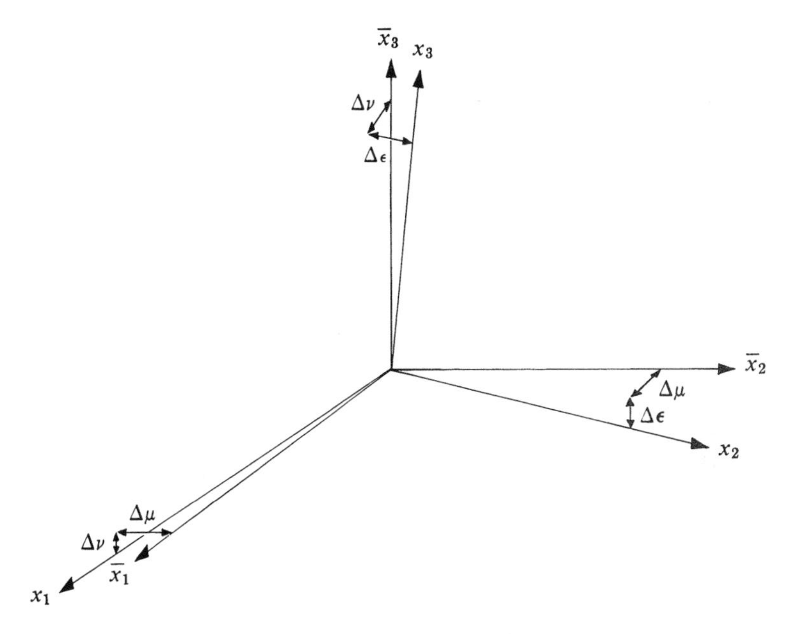

FIGURE 13. *Nutation angles.*

such rotations would not be applied to any satellite positions. They should always be referred, from the start to the end of the computation, to a fixed-coordinate system such that the principal perturbation, the earth's oblateness, is most nearly symmetric with respect to the adopted equator. The equator of this coordinate system should thus be the true equator at an epoch central to the orbital arc being calculated. The equinox can be anything convenient. Some investigators use the equinox defined by the meridian of the equinox at 1950.0; others the true equinox of the orbital epoch. Rotations to coincide with the coordinates used for the orbital arc would have to be applied, however, in two other cases.

1. The right ascension and declination (α, δ) of camera observations are

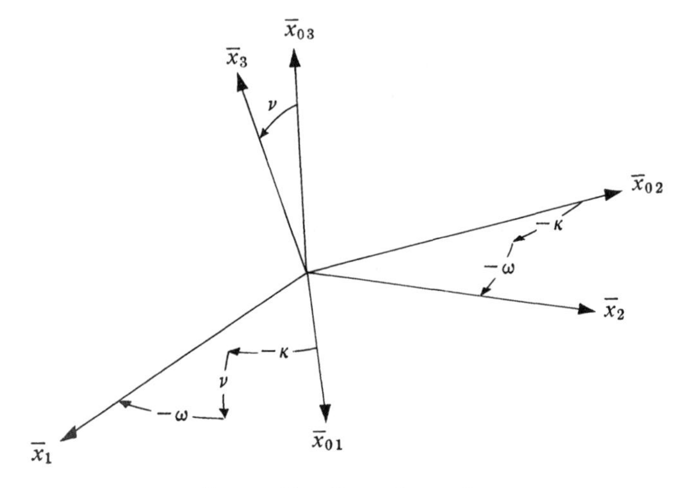

FIGURE 14. *Precession angles.*

normally referred to the coordinate system of the star catalogue or some other standard system, such as the mean equator and equinox of 1950.0. The rotations (4.57) and (4.56) should thus be applied to the unit vector representing the direction $(\bar{\alpha}_0, \bar{\delta}_0)$:

$$\begin{bmatrix} \cos\delta\cos\alpha \\ \cos\delta\sin\alpha \\ \sin\delta \end{bmatrix} = \mathbf{R}_{x\bar{x}}\mathbf{R}_{\bar{x}\bar{x}_0}\begin{bmatrix} \cos\bar{\delta}_0\cos\bar{\alpha}_0 \\ \cos\bar{\delta}_0\sin\bar{\alpha}_0 \\ \sin\bar{\delta}_0 \end{bmatrix}. \tag{4.58}$$

The right ascension α and declination δ equivalent to the direction cosines given by (4.58) can also be obtained directly by (Smart, 1944, pp. 243–244):

$$\alpha = \bar{\alpha}_0 + (\varkappa + \omega) + \Delta\mu + (\nu + \Delta\nu)\sin\alpha\tan\delta - \Delta\epsilon\cos\alpha\tan\delta, \tag{4.59}$$
$$\delta = \bar{\delta}_0 + (\nu + \Delta\nu)\cos\alpha + \Delta\epsilon\sin\alpha.$$

If the orbital positions $\mathbf{x}$ are referred to the true equator and true equinox, (4.59) should be applied as it stands; if the arc is referred to the true equator and some mean equinox, then $\Delta\mu$ should be omitted and the precession $(\varkappa + \omega)$ should be for the time from the epoch to which $\bar{\alpha}_0$ is referred and the epoch of the mean equinox to which $\mathbf{x}$ is referred.

A convenient way of counting time for satellite orbit calculations is the Modified Julian Day, or MJD, in which 1950 Jan 1.0 is 33282.0 and 1960

Jan 0.0 is 36933.0. For time t in MJD, we have the following

$$
\begin{aligned}
T &= (t - 33281.923)/365.242, \\
\Omega_0 &= -0.338(T - 0.61), \\
\lambda_\odot &= 2\pi(T - 0.219), \\
\varkappa &= 0.0001117T + O(10^{-7}), \\
\omega &= 0.0001117T + O(10^{-7}), \\
\nu &= 0.0000969T - O(10^{-7}), \\
\Delta\mu &= -[76.48 \sin \Omega_0 + 5.64 \sin 2\lambda_\odot + O(1.0)] \times 10^{-6}, \\
\Delta\nu &= -[33.3 \sin \Omega_0 + 2.5 \sin 2\lambda_\odot + O(0.4)] \times 10^{-6}, \\
\Delta\epsilon &= [44.7 \cos \Omega_0 + 2.7 \sin 2\lambda_\odot + O(0.4)] \times 10^{-6}.
\end{aligned}
\tag{4.60}
$$

Ω_0 is the longitude of the moon's node and $\lambda_\odot$ is the longitude of the sun. The MJD of 33281.923 is the Besselian year 1950.0 to which mean stellar places are referred.

2. The Greenwich Sidereal Time θ used in the observation equations (4.37)–(4.55) should be the angle between the Greenwich Meridian at the instant of observation and the equinox selected as reference for the orbital positions $\mathbf{x}$. This θ should thus include the same terms resulting from motion of the equinox as would any right ascension on the equator in the same orbital computation. Given a mean sidereal time $\bar{\theta}_{00}$ at an epoch t_{00}, a mean rate of rotation $\dot{\theta}$ with respect to inertial space, and $\mathbf{x}$ coordinates referred to the true equinox at an epoch t_0, we therefore have for time t

$$\bar{\theta} = \bar{\theta}_{00} + \dot{\theta}(t - t_{00}) + (\dot{\varkappa} + \dot{\omega})(t_0 - t_{00}) + \Delta\mu(t_0). \tag{4.61}$$

If the $\mathbf{x}$ coordinates are referred to a mean equinox, then the $\Delta\mu$ term should be omitted.

The time is generally given in days, hours, minutes, and seconds,

$$t = d + h/24 + m/1440 + s/86400. \tag{4.62}$$

Thus, for practical computation a more convenient formula is (taking $\bar{\theta}$ modulo 2π)

$$
\begin{aligned}
\bar{\theta} = \bar{\theta}_{00} &+ (\dot{\theta} + \dot{\varkappa} + \dot{\omega} - 2\pi)(t - t_{00}) + 2\pi(h/24 + m/1440 + s/86400) \\
&+ (\dot{\varkappa} + \dot{\omega})(t_0 - t) + \Delta\mu(t_0).
\end{aligned}
\tag{4.63}
$$

Some numerical values are

$$\begin{aligned} t_{00} &= 36933.0, \\ \bar{\theta}_{00} &= 1.72218613, \\ (\dot{\theta} + \dot{\varkappa} + \dot{\omega} - 2\pi) &= 0.0172027913, \\ \dot{\varkappa} + \dot{\omega} &= 0.611 \times 10^{-6}. \end{aligned} \tag{4.64}$$

The overbar on the sidereal time $\bar{\theta}$ in (4.61) and (4.63) signifies that these equations are based on the assumption that the earth rotates uniformly about an axis fixed with respect to the crust to which the tracking stations of coordinates $\mathbf{u}_0$ in the observation equations (4.37)–(4.55) are also fixed. This assumption is incorrect: shifts of mass in the earth's atmosphere cause variations in the polar axis of about 5 meters in position and in the time of about 0.03^s. The displacement along the equator of $0.03^s\dot{\theta}a_e$ is about 14 meters so if we are interested in accuracies of this sort, the Greenwich sidereal time must be further corrected beyond (4.63) for use in (4.37)–(4.55).

Connected with the shifts in the earth's axis and the changes in rotation rate are three types of time defined by the earth's rotation, called Universal time (UT):

1. UT0 refers to the instantaneous rotation about the instantaneous axis: it is what would be derived from observations of transits of stars across the meridian of an earth-fixed observatory.

2. UT1 refers to the instantaneous rotation about the mean axis, as defined by averaging over 6 years' observations by the International Latitude Service. (Six years is chosen as being about the lowest common multiple of the forced annual and the free 14-month variation in latitude.)

3. UT2 refers to an approximation of the mean rotation about the mean axis: it differs from UT1 by an estimated seasonal variation of about $0^s.03$.

In correcting the Greenwich Sidereal Time we are concerned only with a rotation about the 3-axis. Hence, UT1 is the appropriate time to use with satellite observations. If the orbit is being calculated in Atomic Time A1, we therefore have a final correction $\Delta\theta$ to apply to the mean $\bar{\theta}$ from (4.63). Thus,

$$\theta = \bar{\theta} + \Delta\theta = \bar{\theta} + \dot{\theta}(\text{UT1} - \text{A1}). \tag{4.65}$$

Since UT1 is an observationally determined quantity, a formula cannot be given for the difference (UT1-A1). Time signal bulletins giving the differences between UT2, UT1, UT0, A1 and emitted signals are issued quarterly by the Royal Greenwich Observatory and the United States Naval Observatory. The magnitude of UT1-A1 is about $2\frac{1}{2}^s$.

4.7. Observability Conditions

There are several situations in which it is necessary to determine whether, for a given combination of tracking station and satellite position, the satellite is observable. The most obvious condition is that the satellite is above the horizon of the tracking station:

$$l_3 > 0, \tag{4.66}$$

where l_3 is the vertical coordinate of the $\mathbf{l}_T$ system defined by (4.17). From (4.17)

$$l_3 = \mathbf{k}\mathbf{R}_{lu}[\mathbf{R}_3(\theta)\mathbf{R}_{xq}\mathbf{q} - \mathbf{u}_0] > 0, \tag{4.67}$$

where $\mathbf{k}$ is the transpose $\{0,\ 0,\ 1\}$ of the unit vector along the 3-axis. $\mathbf{k}\mathbf{R}_{lu}\mathbf{u}_0$ is approximately the radial distance of the station from the earth's center. $\mathbf{R}_{lu}$ is constant with time and $\mathbf{R}_{xq}$, defined by (2.32), varies slowly with motion of perigee and node. Thus, for a given day we can write (4.67) rather accurately as

$$[r_{31},\ r_{32},\ r_{33}]\begin{bmatrix} \cos\theta & \sin\theta & 0 \\ -\sin\theta & \cos\theta & 0 \\ 0 & 0 & 1 \end{bmatrix}\begin{bmatrix} s_{11} & s_{12} \\ s_{21} & s_{22} \\ s_{31} & s_{32} \end{bmatrix}\begin{bmatrix} a(\cos E - e) \\ a\sqrt{1 - e^2}\sin E \end{bmatrix} - R_E > 0, \tag{4.68}$$

where the r_{ij}'s are elements of $\mathbf{R}_{lu}$, $\mathbf{R}_3(\theta)$ has been defined by (2.7), the s_{ij}'s are elements of $\mathbf{R}_{xq}$, $\mathbf{q}$ has been defined by (3.23), and R_E is the radius of the earth. Multiplying out (4.68), we get for horizon intersection

$$f(\theta)\cos E + g(\theta)\sin E - h(\theta) = 0, \tag{4.69}$$

where

$$\begin{aligned} f(\theta) &= a[(r_{31}\cos\theta - r_{32}\sin\theta)s_{11} + (r_{31}\sin\theta + r_{32}\cos\theta)s_{21} + r_{33}s_{31}], \\ g(\theta) &= a\sqrt{1 - e^2}\,[(r_{31}\cos\theta - r_{32}\sin\theta)s_{12} + (r_{31}\sin\theta + r_{32}\cos\theta)s_{22} + r_{33}s_{32}], \\ h(\theta) &= R_E + ef(\theta). \end{aligned} \tag{4.70}$$

If we substitute $\sqrt{1 - \cos^2 E}$ for $\sin E$, solve the resulting quadratic equation for $\cos E$; and make a similar substitution for $\cos E$ and solution for $\sin E$, in order to obtain the eccentric anomaly E without quadrant ambiguity, we get

$$\cos E = \frac{fh \pm g\sqrt{f^2 + g^2 - h^2}}{f^2 + g^2},$$
$$\sin E = \frac{gh \mp f\sqrt{f^2 + g^2 - h^2}}{f^2 + g^2}. \tag{4.71}$$

The condition for observability will be that the solutions are real, that is,

$$f^2 + g^2 > h^2. \tag{4.72}$$

For any particular revolution $0 < M < 2\pi$ of a satellite, we can write for the Greenwich Sidereal Time θ

$$\theta = \theta_0 + \frac{\dot{\theta}}{\dot{M}} M, \tag{4.73}$$

where θ_0 is the sidereal time at perigee and $\dot{M}$ is calculated by (3.20) plus (3.74). For the combination of a particular station, defining $\mathbf{R}_{lu}$; a particular day, defining $\mathbf{R}_{xq}$; and a particular revolution, defining θ_0, we can iteratively determine whether observability will occur by starting with the approximation

$$(1) \quad \theta = \theta_0 + \frac{\dot{\theta}}{\dot{M}} \pi; \tag{4.74}$$

then we

(2) calculate f, g, h by (4.70);
(3) apply the test (4.72);
(4) if (4.72) test passed, solve (4.71) for the two roots E_1, E_2;
(5) recalculate θ, from (3.19), by

$$\theta = \theta_0 + \frac{\dot{\theta}}{2\dot{M}} [E_1 + E_2 - e(\sin E_1 + \sin E_2)]; \tag{4.75}$$

and (6) return to step (2), until the changes in E_1 and E_2 become insignificant.

For the intersection of the earth's shadow by the satellite orbit, a similar sort of iterative scheme can be set up. Let the $\mathbf{s}$ coordinate system have a geocentric origin with the s_1 axis toward the sun. Then the condition of shadow intersection can be written as (see Figure 15)

$$s_1 < 0,$$
$$s_2^2 + s_3^2 = R_E^2, \tag{4.76}$$

or

$$s_1 = -\sqrt{r^2 - R_E^2}. \tag{4.77}$$

The sun-referred coordinates **s** can be related to the equator-equinox referred coordinates **x** by the elements of the sun's orbit referred to the earth. Since this orbit defines the equinox, its nodal longitude Ω is always zero. Then, from Figure 3

$$\mathbf{R}_{sx} = \mathbf{R}_3(\omega + f)\mathbf{R}_1(\epsilon) = \mathbf{R}_3(\lambda_\odot)\mathbf{R}_1(\epsilon), \tag{4.78}$$

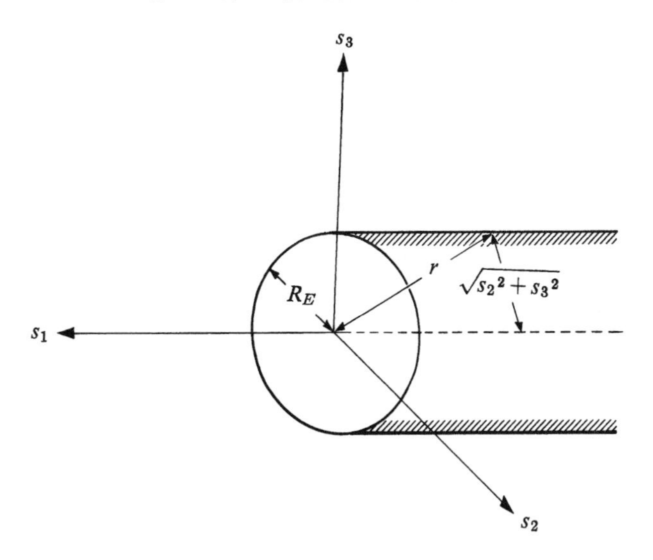

FIGURE 15. *Orbit and shadow relationship.*

where $\lambda_\odot$ is the longitude of the sun, from (4.60), and ϵ is the obliquity, 23°27′. Then (4.77) can be written entirely in terms of the sun's and the satellite's orbits. Thus

$$s_1 = \mathbf{i}\mathbf{R}_{sx}\mathbf{R}_{xq}\mathbf{q} = -\sqrt{r^2 - R_E^2}\,, \tag{4.79}$$

where **i** is the transpose $\{1, 0, 0\}$ of the unit vector along the 1-axis, or

$$r_{11}a(\cos E - e) + r_{12}a\sqrt{1 - e^2}\sin E = -\sqrt{a^2(1 - e\cos E)^2 - R_E^2}, \tag{4.80}$$

where r_{11}, r_{12} are elements of $\mathbf{R}_{sx}\mathbf{R}_{xq}$, $\mathbf{q}$ has been defined by (3.23), and r has been defined by (3.15). (4.80) can be arranged in a manner similar to (4.69):

$$f\cos E + g\sin E - h(E) = 0, \tag{4.81}$$

where we have

$$f = r_{11}a,$$
$$g = r_{12}a\sqrt{1-e^2}, \qquad (4.82)$$
$$h(E) = r_{11}ae - \sqrt{a^2(1 - e\cos E)^2 - R_E^2}.$$

The differences from (4.69)–(4.70) are that now f, g vary much more slowly—being functions of $\lambda_\odot$, ω, Ω—while $h(E)$ has a rapid but moderate variation because it is a function of the anomaly E, or M, through a term with a small multiplier, the eccentricity e. Hence an iterative solution using (4.71) and (4.72) can again be applied. To be sure to pass the test (4.72), the initial value of $h(E)$ should be one to give a minimum value of h^2; namely, we should

(1) assume $\cos E$ is 1;
(2) calculate f, g, h by (4.82);
(3) apply the test (4.72);
(4) if (4.72) test passed, solve (4.71) for the two roots E_1, E_2;
(5) for each root, recalculate $h(E_1)$, $h(E_2)$ by (4.82); and
(6) return to step (2), iterating separately for E_1 and E_2 until the changes become insignificant.

Horizon and shadow intersection can also be solved by graphical means; see Veis (1961, 1963b).

The term "observability" must, in addition to the geometrical conditions we have discussed so far, also include the problems of signal strength and atmospheric attenuation, the latter of which becomes prohibitively severe, of course, if we try to use a camera in cloudy weather.

REFERENCES

1. Brown, D. C. "An Advanced Reduction and Calibration for Photogrammetric Cameras." *AF Cambridge Res. Lab. Tech. Rep. 64–40* (1964), 113 pp.

2. Gaposchkin, E. M. "Differential Orbit Improvement (DOI-3)." *Smithsonian Inst. Astr. Obs. Spec. Rep. 161*, 1964.

3. Guier, W. H. "Studies on Doppler Residuals—I: Dependence on Satellite Orbit Error and Station Position Error." *Johns Hopkins Univ. Appl. Phys. Lab. Rep TG-503* (1963a), 72 pp.

4. Guier, W. H. "Ionospheric Contributions to the Doppler Shift at VHF from Near Earth Satellites." *Johns Hopkins Univ. Appl. Phys. Lab. Rep. CM-1040* (1963b), 53 pp.

5. Hopfield, H. S. "The Effect of Tropospheric Refraction on the Doppler Shift of a Satellite Signal." *J. Geophys. Res.*, *68* (1963), pp. 5157–5168.

6. Kaula, W. M. "Analysis of Gravitational and Geometric Aspects of Geodetic Utilization of Satellites." *Geophys. J.*, 5 (1961), pp. 104–133.

7. Kaula, W. M. "Celestial Geodesy." *Advan. Geophys.*, *9* (1962), pp. 191–293.

8. Mueller, I. I. *Introduction to Satellite Geodesy.* New York: Frederick Ungar Publishing Company Inc., 1964.

9. Nautical Almanac Offices of the United Kingdom and the United States of America. *Explanatory Supplement to the American Ephemeris and Nautical Almanac.* London: H. M. Stationery Office, 1961.

10. Smart, W. M. *Spherical Astronomy.* 4th ed. London: Cambridge University Press, 1944.

11. Veis, G. "Geodetic Uses of Satellites." *Smithsonian Contrib. to Astrophys.*, *3* (1960), pp. 95–161.

12. Veis, G. (Ed.). *The Use of Artificial Satellites for Geodesy.* Amsterdam: North-Holland Publishing Company, 1963a.

13. Veis, G. "Optical Tracking of Artificial Satellites," *Space Sci. Rev.*, *2* (1963b), pp. 250–296.

5

STATISTICAL IMPLICATIONS

5.1. General

In Chapter 3 we discussed the variation of the close satellite orbit with respect to time t as expressed by the six quantities $\{\mathbf{x}, \dot{\mathbf{x}}\}$ or $\{a, e, i, M, \omega, \Omega\}$. These variations can be expressed as periodic oscillations superimposed on a secularly changing model: namely, a 6-dimensional vector time series. The subject of the statistical study of time series has received considerable attention in recent years, so we should be able to apply some of the results of this study to the geodetic satellite problem. In the last section of Chapter 4, however, it was suggested that there might be a special difficulty treating the satellite as a time series in that there are definite geometrical limitations on when observations can be made.

In Chapter 4 we discussed the formation of observation equations which included two types of corrections: corrections to parameters dP_j and corrections to observations dO_i. The solution of sets of such equations by the method of least squares and its generalization is also a subject that has received considerable attention. There was also indicated, however, a special difficulty. The so-called corrections to observations dO_i necessarily absorb a lot of the discrepancy between the mathematical model and actuality which does not have the character of randomness of errors expected of well-programmed instrumental observations. We hope these two statistical ways of viewing a satellite orbit and observations thereof can be combined effectively.

5.2. Time Series

Let $\mathbf{y}(t)$ be a continuous function of time that may be a vector of any number of dimensions. For any duration of time T it can evidently be represented to any desired degree of accuracy as a sum of sinusoidal terms.

Thus we have

$$\mathbf{y}(t) = \sum_{n=0}^{\infty}\left[\mathbf{a}_n \cos\frac{2\pi n}{T}\,t + \mathbf{b}_n \sin\frac{2\pi n}{T}\,t\right], \qquad 0 \le t \le T, \tag{5.1}$$

where, given $\mathbf{y}(t)$ for $0 \le t \le T$,

$$\begin{aligned} \mathbf{a}_n &= \frac{2-\delta_{on}}{T}\int_0^T \mathbf{y}(t)\cos\frac{2\pi n}{T}\,t\,dt \\ \mathbf{b}_n &= \frac{2}{T}\int_0^T \mathbf{y}(t)\sin\frac{2\pi n}{T}\,t\,dt. \end{aligned} \tag{5.2}$$

Since (5.2) is a consequence of the orthogonality property,

$$\int_0^T \begin{bmatrix}\sin\\ \text{or}\\ \cos\end{bmatrix}\frac{2\pi n}{T}\,t \begin{bmatrix}\sin\\ \text{or}\\ \cos\end{bmatrix}\frac{2\pi m}{T}\,t\,dt = 0, \qquad m \ne n. \tag{5.3}$$

If $\mathbf{y}(t)$ is a normal, "well-behaved" function, staying within certain bounds, the coefficients $\mathbf{a}_n$, $\mathbf{b}_n$ will, in general, decrease in magnitude as n increases; namely,

$$\lim_{n\to\infty}\begin{bmatrix}\mathbf{a}_n\\ \mathbf{b}_n\end{bmatrix} = 0. \tag{5.4}$$

As the duration T is lengthened, if $\mathbf{y}$ stays within the same bounds, the Fourier representation (5.1) remains feasible, and if the quadratic sum of coefficients $\mathbf{a}_n$, $\mathbf{b}_n$ representing the amount of variability within a certain frequency band Δf approaches a constant value, then

$$\lim_{T\to\infty}\mathbf{C}(f,\ \Delta f,\ T) = \text{const}, \tag{5.5}$$

where for each element C_i of $\mathbf{C}$ corresponding to elements a_{in}, b_{in} of $\mathbf{a}_n$, $\mathbf{b}_n$,

$$C_i(f,\ \Delta f,\ T) = \sum_n (a_{in}^2 + b_{in}^2), \qquad T\left[f - \frac{\Delta f}{2}\right] \le n \le T\left[f + \frac{\Delta f}{2}\right]. \tag{5.6}$$

Hence, in this case, we can express $\mathbf{y}(t)$ in the limit as

$$\mathbf{y}_p(t) = \sum_{k=1}^{l}[\mathbf{c}_k\cos 2\pi f_k t + \mathbf{d}_k \sin 2\pi f_k t] + \int_0^\infty [\mathbf{g}(f)\cos 2\pi f t + \mathbf{h}(f)\sin 2\pi f t]\,df, \tag{5.7}$$

where the subscript p denotes periodic, and where the summation is over a finite set of discrete frequencies (called a line spectrum) and the integration is over a continuous variation of density $\mathbf{g}(f)$, $\mathbf{h}(f)$ with respect to frequency (called a continuous spectrum). Evidently, for Δf small enough, an element C_i of $\mathbf{C}(f_k, \Delta f, \infty)$ is $(c_{ik}^2 + d_{ik}^2)$.

However, it may be the case that the condition (5.5) does not hold true for a representation of $\mathbf{y}(t)$ by the Fourier series (5.1) because $\mathbf{y}(t)$ does not remain within certain bounds, but instead increases or decreases or oscillates more and more widely with time. In this case, $\mathbf{y}(t)$ has to be expressed as

$$\mathbf{y}(t) = \mathbf{y}_s(t) + \mathbf{y}_p(t), \tag{5.8}$$

where $\mathbf{y}_p(t)$ is gotten from (5.7) and where

$$\mathbf{y}_s(t) = \sum_{j=1}^{\infty} \mathbf{z}_j(t - t_j)^j. \tag{5.9}$$

The subscript s denotes secular. In this case, for a sufficiently long duration T, the dominant change in $\mathbf{y}$ will always be expressed by the secular part $\mathbf{y}_s$. In a particular application, the lower $\mathbf{z}_j$, $\mathbf{z}_1$ and $\mathbf{z}_2$ may be appreciably smaller than the lower $\mathbf{a}_n$, $\mathbf{b}_n$, and the $\mathbf{z}_j$ may go to zero rapidly with increase in j, so that over short and moderate durations T it will still be feasible to study $\mathbf{y}$ as the periodic function $\mathbf{y}_p$. If the $\mathbf{z}_j$ are known, this is very simply done by just subtracting out $\mathbf{y}_s$. However, if the $\mathbf{z}_j$ are not known, then there will always be some distortion of the $\mathbf{a}_n$, $\mathbf{b}_n$ because of the lack of orthogonality, such as (5.3), with the $\mathbf{z}_j$.

$\mathbf{y}_p$ is known as a stationary function, because its statistical properties are constant with time. The leading statistical property of $\mathbf{y}_p$ is its mean square. Since $\mathbf{y}_p$ is not necessarily a position vector and since its different components may even have different dimensions, it is appropriate to consider this mean square as a vector itself (rather than a dot product), of which each component y_i^2 is

$$\sigma^2[y_i] = E[y_i^2] = \frac{1}{2}\left\{\sum_{k=1}^{l}[c_{ik}^2 + d_{ik}^2] + \int_0^{\infty}\int_0^{\infty}[g_i^2(f) + h_i^2(f)]\,df\,df\right\}. \tag{5.10}$$

Equation (5.10) is a consequence of the orthogonality property (5.3) with T extended to approach infinity. (5.10) can be generalized in two ways. First, there will in general be a nonzero mean product $E\{y_iy_j\}$ (called cross-variance), second, and a nonzero mean product $E\{y_i(t)y_i(t + \tau)\}$ (called covariance). The cross-variance will generally be a known function of the variance—given the c_{ik}^2, d_{ik}^2, g_i^2, h_i^2 for one component, they can be calculated for the others—but this is not in general true of the covariance.

Multiplying $y_{ip}(t)$ by $y_{ip}(t + \tau)$ from (5.7), converting products of sines and cosines to sums, and applying (5.3), we get

$$\begin{aligned}\mathrm{Cov}_i\,(\tau) &= E[y_i(t)y_i(t + \tau)] \\ &= \frac{1}{2}\left\{\sum_{k=1}^{l}[(c_{ik}^2 + d_{ik}^2)\cos 2\pi f_k\tau \right. \\ &\quad \left. + \int_0^\infty \int_0^\infty [g_i^2(f) + h_i^2(f)]\cos 2\pi f\tau\, df\, df\right\}. \end{aligned} \tag{5.11}$$

The osculating Keplerian elements $\{a, e, i, M, \omega, \Omega\}$ of a close satellite orbit can be expressed by an appropriate combination of (3.113), (3.76), and (3.153)–(3.158). Most—say more than 99%—of the variation with time can be regarded as known, including the dominant secular terms $\dot{M}$, $\dot{\Omega}$, $\dot{\omega}$ arising from the central term kM and J_2 and $\dot{a}$, $\dot{e}$, di/dt, $\ddot{M}$, $\ddot{\Omega}$, $\ddot{\omega}$ arising from simple models of the atmosphere and radiation pressure. The residuals of the actual motion with respect to the known part are what we are interested in examining as a vector time series $\mathbf{y}(t)$. This time series will still contain significant $\dot{a}$, $\dot{e}$, di/dt, $\ddot{M}$, $\ddot{\Omega}$, $\ddot{\omega}$ as a result of inadequacies of the atmospheric model. However, these six terms can be reduced to two by expressing the others as functions of the acceleration in the mean anomaly and the rate of rotation of the atmosphere. For durations of record T of satellites of interest to geodesy, the irregular balance of drag effects can be represented by the continuous spectrum in (5.7). To compare this drag spectrum to that arising from the variations of the gravitational field, we need to calculate the order-of-magnitude of the accelerations involved. For drag, take as typical $C_D = 2.4$, $A/m = 0.05$ cm²/gm, $v = 7.5$ km/sec, and as pessimistic $\rho \leq 10^{-14}$ gm/cm³. (See the table of numerical values). Using these values in the Equation (3.152) for the drag force obtains for an estimate of the drag acceleration:

$$\sigma\{a_d\} \leq \pm 2.4 \times 0.05 \times 10^{-14} \times (7.5 \times 10^5)^2/2 \approx \pm 3 \times 10^{-4} \text{ cm/sec}^2.$$

For an estimate of the acceleration due to a variation of the gravitational field, take the radial derivative of the potential term V_{22} as given by (3.53) for a semimajor axis of 8×10^3 km and a magnitude 1.5×10^{-6} for J_{22}:

$$\begin{aligned}\sigma(a_g\} &\approx \pm 3.986 \times 10^{20} \times (6.378 \times 10^8)^2 \times 1.5 \times 10^{-6}/(8 \times 10^8)^4 \\ &\approx \pm 6 \times 10^{-4} \text{ cm/sec}^2.\end{aligned}$$

Thus even for a satellite orbit far from ideal—atmospheric densities of 10^{-14} gm/cm³ would exist only at altitudes below 600 km on the sunlit side near the peak of the solar cycle—the acceleration due to the variations in the gravitational field would be comparable to that due to drag. Furthermore, the drag acceleration is spread out into a continuous spectrum, whereas, according to the transformation (3.70), the spectrum of V_{lm} is, for a small eccentricity, comprised almost entirely by only $(l + 1)$ lines: for V_{22}, three lines, one near a frequency of two cycles/day and two near two cycles/revolution.

For a length of record T, a continuous spectrum of amplitude density $\mathbf{g}(f_k)$ will obscure a discrete term of amplitude less than $\mathbf{c}_k = \mathbf{g}(f_k)/T$, as indicated by (5.1)–(5.2). If we assume that the mean square drag acceleration $\sigma^2\{a_d\}$ is distributed uniformly from one cycle per revolution, $1/P$, downward in frequency, then the density $g^2(f_k)$ will be $P\sigma^2\{a_d\}$. For a length of record T the effective number of frequency bands will be T/P and the variability $\sigma_f^2\{a_d\}$ in a particular band will be $P\sigma^2\{a_d\}/T$. Hence the record T would not have to be very long for the gravitational acceleration at a particular frequency to stand out above the drag. Therefore

$$c_k \approx \sigma\{a_g\}/\sqrt{l+1} \gg \sigma\{a_d\}\sqrt{P/T}.$$

Since accelerations are not observed, to compare variability at different frequencies it is better to integrate to obtain the spectrum of position variation:

$$\sigma_f\{\Delta s\} = \sigma_f\{a\}/(2\pi f)^2.$$

To compare the implications of the numerical estimates of $\pm 3 \times 10^{-4}$ cm/sec² for gravitational acceleration $\sigma\{a_g\}$, assuming the equipartition below $1/P$ for $\sigma^2\{a_d\}$, we take a period P of 96 minutes and a record T of 3 months. Then

$$\begin{aligned}\sigma_f\{\Delta s_d\} &= \sigma\{a_d\}\sqrt{P/T}\big/(2\pi f)^2 \\ &= \pm 3 \times 10^{-4}\sqrt{1/15 \times 90}\big/(2\pi f/86{,}400)^2 = \pm 15/f^2 \text{ meters}\end{aligned}$$

for frequency f in cycles per day. For a frequency of 2 cycles per day, the perturbation is thus about ± 4 meters, compared to ± 165 meters obtained from a gravitational acceleration of $\pm 6 \times 10^{-4}/\sqrt{3}$ cm/sec² at the same frequency. However, for a lower frequency the drag perturbation will obviously become much greater—for example, for one cycle/month, the above estimate gives ± 13.5 kilometers. The situation is shown schematically in Figure 16, which would be characteristic of a perigee height around 600 km for a length of record of about 3 months. For any type of effect, the spectrum rises steeply toward the lower frequency end, due to the longer integration time. Even though they may have considerably less amplitude than drag effects of lower frequency, the orthogonality condition (5.3) still enables accurate determination of the discrete high frequency perturbations caused by tesseral harmonics—the V_{lm} for which $m \neq 0$—provided that $\mathbf{y}(t)$ is completely and continuously observed. However, as indicated in Chapter 4, observations are incomplete in the sense that only one or two components out of the six comprising **y** are measured: a range or range-rate or photo plate coordinates. They are also discontinuous because of the geometrical limitations on observability discussed in Section 4.7. Consequently, the time series we are forced to consider is not $\mathbf{y}(t)$, but instead a linear transformation thereof multiplied by a "window" function. In addition, this time series will have a part arising from other sources than **y**,

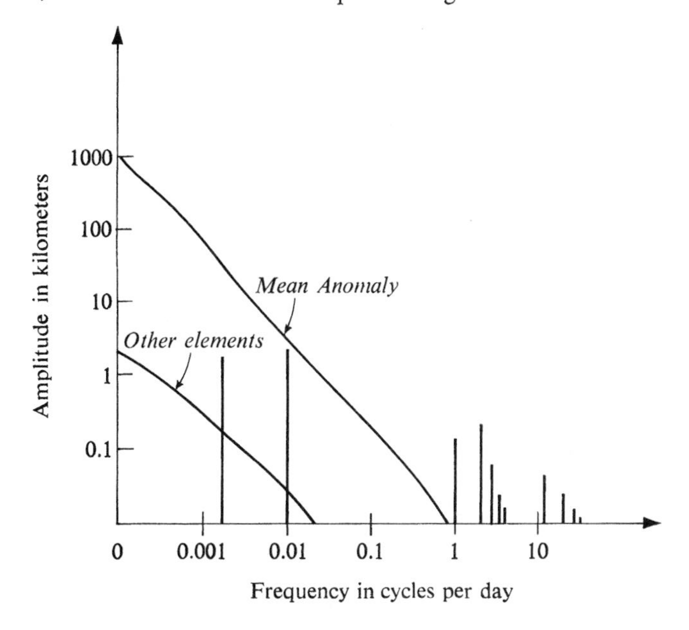

FIGURE 16. *Spectrum of satellite orbit variations.*

due to errors in station position, and so forth:

$$\mathbf{w}(t) = \frac{\partial \mathbf{w}}{\partial \mathbf{y}} \mathbf{y} I(t) + \frac{\partial \mathbf{w}}{\partial \mathbf{u}} \mathbf{u}. \tag{5.12}$$

The Jacobian $\partial \mathbf{w}/\partial \mathbf{y}$ is constituted by the partial derivatives of the observations with respect to the osculating elements, as developed in (4.37)–(4.51), and is itself a function of time. The window function $I(t)$ is unity during observation and zero at other times. The station error function **u** will be equal to the error of the station position during observation and zero at other times.

$\partial \mathbf{w}/\partial \mathbf{y}$, $I(t)$, and $\partial \mathbf{w}/\partial \mathbf{u}$ all have complicated spectrums, predominantly of high frequency, comparable to that of the satellite orbit itself. Consequently contributions to $\mathbf{w}(t)$ from the discrete gravitationally caused spectrum will be difficult to distinguish from those from the continuous drag spectrum. Simple harmonic analysis will not suffice; instead, some sort of least-squares or quadratic sum minimization is necessary. Since such a procedure involves squares and products, we are still interested in the statistical properties of the orbital time series, expressed by the variances (5.10) and covariances (5.11).

The discrete variances $\{c_{ik}^2 + d_{ik}^2\}/2$ arising from variations in the earth's gravitational field can be calculated from estimates of the variances of the spherical harmonic coefficients C_{lm}, S_{lm} of the gravitational field. These estimates are best obtained from autocovariance analysis of gravimetry, which estimates the degree variances σ_l^2 defined by

$$\sigma_l^2 = \Sigma_m(\bar{C}_{lm}^2 + \bar{S}_{lm}^2), \tag{5.13}$$

where $\bar{C}_{lm}$, $\bar{S}_{lm}$ are the coefficients of the normalized spherical harmonics defined by (1.34). A rough rule for the σ_l^2 determined from autocovariance analysis of gravimetry is

$$\sigma_l^2 \approx 160 \times 10^{-12}/l^3. \tag{5.14}$$

From (5.13) and (5.14),

$$\sigma\{\bar{C}_{lm}, \bar{S}_{lm}\} = \frac{\sigma_l}{\sqrt{2l+1}} \approx \pm \frac{10^{-5}}{l^2}. \tag{5.15}$$

A given frequency f_k in (5.7) will, from (3.71), correspond to a set of particular combinations of the indices l, m, p, q in the orbitally referred expression of the spherical harmonics. Thus

$$f_k = [(l-2p)\dot{\omega} + (l-2p+q)\dot{M} + m(\dot{\Omega} - \dot{\theta})]/2\pi. \tag{5.16}$$

The lowest degree term giving rise to a particular frequency f_k will be that for which

$$l = m \qquad \text{unless } m \text{ is 0 or 1}. \tag{5.17}$$

Others in the set will have

$$\begin{aligned} l &= m + 2j, \\ p &= p_j = p_0 + j, \end{aligned} \tag{5.18}$$

where j is any integer. Consequently, from (5.15) and (4.27),

$$E\left[\frac{1}{2}(c_{ik}^2 + d_{ik}^2)\right] = \sum_{j=0}^{\infty} \frac{\sigma_l^2}{2l+1} K_{ilmpq}^2, \tag{5.19}$$

where l and p are obtained from (5.18) and K_{ilmpq} is the coefficient of the partial derivative of the Keplerian element with respect to C_{lm}, as given by (4.28).

For estimates of the spectral representation of the drag effects $\frac{1}{2}[\mathbf{g}^2(f) + \mathbf{h}^2(f)]$, analysis must be made of orbital residuals. Here the dominant effect will be in the mean anomaly, so much so that the inexactness of the present data makes it not worthwhile to consider the errors in other elements caused by drag. The results of analysis of orbits for drag effects are usually expressed in terms of the rate of change of period $\dot{P}$. Autocovariance analyses of $\dot{P}$ generally obtain an exponential drop-off with increasing frequency for its power spectrum:

$$\sigma_f^2\{\dot{P}\} = D\langle\dot{P}\rangle^2 \exp(-qf). \tag{5.20}$$

where $\langle\dot{P}\rangle$ is the mean value of the rate-of-change of period.

The relationship between sinusoidal oscillations in $\dot{P}$ and in $\ddot{M}$ will be

$$\Delta\ddot{M} = -\frac{n^2}{2\pi}\Delta\dot{P} = -\frac{n^2}{2\pi}\cos 2\pi ft. \tag{5.21}$$

Solving for $\sigma_f^2\{M\}$ by integrating twice and squaring, we get

$$\sigma_f^2\{M\} = \frac{n^4}{64\pi^6 f^4} D\langle\dot{P}\rangle^2 \exp(-qf). \tag{5.22}$$

For f in cycles per day, numerical estimates are 0.6 for D and 19.2 for q. The practical implication of the frequency f in the denominator is that some device must always be applied to absorb low frequency drag effect—usually both arbitrary polynomials in the mean anomaly and limiting the duration of an orbital arc. If h polynomials are used for an arc of duration W, the lower limit of the unabsorbed frequencies will be about $h/2W$.

As previously pointed out, the correction to observation dO_i of the observation equation (4.32) must absorb all discrepancy between the mathematical model and actuality. Hence the contribution to the variance of dO_i, $\sigma^2\{O_i\}$ from neglected coefficient C_{lm}, S_{lm} will, from (5.19), be

$$\Delta_{lm}\sigma^2\{O_i\} = \sum_{p,q}\left[\frac{\partial O_i}{\partial s_h} K_{hlmpq} \frac{\sigma_l^2}{2l+1} K_{klmpq} \frac{\partial O_i}{\partial s_k}\right], \tag{5.23}$$

where (s_h, s_k) are either both $(a, e, \text{or } i)$ or both $(M, \omega, \text{or } \Omega)$; contributions from cross products are zero. The corresponding contribution to the covariance of dO_i and dO_j is

$$\Delta_{lm}\,\mathrm{Cov}\,\{O_i O_j\} = \sum_{p,q}\left[\frac{\partial O_i}{\partial s_h} K_{hlmpq} \frac{\sigma_l^2}{2l+1} K_{klmpq} \frac{\partial O_j}{\partial s_k}\right]$$
$$\times \cos[(l-2p)(\omega_i - \omega_j) + (l-2p+q)(M_i - M_j) + m(\Omega_i - \Omega_j - \theta_i + \theta_j)]. \tag{5.24}$$

Similarly, for the drag effect, we have

$$\Delta_d\sigma^2\{O_i\} = \left[\frac{\partial O_i}{\partial M}\right]^2 \int_{h/2W}^{\infty} \sigma_f^2\{M\}\, df, \tag{5.25}$$

$$\Delta_d\,\mathrm{Cov}\,\{O_i, O_j\} = \frac{\partial O_i}{\partial M}\frac{\partial O_j}{\partial M} \int_{h/2W}^{\infty} \sigma_f^2\{M\} \cos 2\pi f(t_i - t_j)\, df. \tag{5.26}$$

5.3. Quadratic Sum Minimization

We have set up in (4.32) a standard form for an observation equation, which states that the discrepancy between an observed quantity and a calculated quantity based on a mathematical model must be accounted for by a combination of corrections to the parameters on which the mathematical model is based and a correction to the observation. Normally, we must deal with a set of such observation equations that is much larger than the number of model parameters that are to be corrected. Let us write a set of observation equations (4.32) in matrix form. Thus

$$\underset{n\times 1}{-\mathbf{x}} + \underset{n\times m}{\mathbf{M}}\ \underset{m\times 1}{\mathbf{z}} = \underset{n\times 1}{\mathbf{f}}, \tag{5.27}$$

where $m < n$, and

$$\mathbf{x} = \begin{bmatrix} dO_1 \\ \cdot \\ \cdot \\ \cdot \\ dO_i \\ \cdot \\ \cdot \\ \cdot \\ dO_n \end{bmatrix}, \qquad \mathbf{M} = \begin{bmatrix} \partial C_1/\partial P_1 & \partial C_1/\partial P_2 & \cdots & \partial C_1/\partial P_m \\ \partial C_2/\partial P_1 & & & \\ & \ddots & & \vdots \\ \vdots & & \partial C_i/\partial P_j & \vdots \\ & & \ddots & \\ \partial C_n/\partial P_1 & & \cdots & \partial C_n/\partial P_m \end{bmatrix},$$

$$\mathbf{z} = \begin{bmatrix} dP_1 \\ dP_2 \\ \cdot \\ \cdot \\ \cdot \\ dP_j \\ \cdot \\ \cdot \\ \cdot \\ dP_m \end{bmatrix}, \qquad \mathbf{f} = \begin{bmatrix} O_1 - C_1 \\ O_2 - C_2 \\ \cdot \\ \cdot \\ \cdot \\ O_i - C_i \\ \cdot \\ \cdot \\ \cdot \\ O_n - C_n \end{bmatrix}. \tag{5.28}$$

The derivation of (4.32), and of the other equations in Chapter 4 pertaining to observation equations, are based on the assumption that the corrections were differentials. This assumption is equivalent to assuming that the

$dP_j dP_k$ term in a Taylor series development for $(O_i + dO_i - C_i)$ is negligible:

$$(O_i + dO_i - C_i) = \sum_j \frac{\partial C_i}{\partial P_j} dP_j + \frac{1}{2} \sum_{j,k} \frac{\partial C_i}{\partial P_j} \cdot \frac{\partial C_i}{\partial P_k} dP_j \, dP_k + \cdots, \tag{5.29}$$

which in turn implies that $\partial C_i / \partial P_j$ in (5.29), and hence $\mathbf{M}$ in (5.27), can be considered constant. (5.27) is thus a set of linear equations. The vector $\mathbf{f}$ can be considered as the coordinates of a point in n-dimensional space, and the vector $\mathbf{x}$ as corrections to these coordinates. Hence

$$\mathbf{Mz} = \mathbf{f} + \mathbf{x} \tag{5.30}$$

is a set of equations that together represent a linear form in the n-dimensional space, since we obviously can express the m elements of $\mathbf{z}$ in terms of the n elements of $\mathbf{f} + \mathbf{x}$ by selecting and solving a set of m equations from (5.27). For example, if n is equal to 3 and m is equal to 1, (5.27) would be

$$\begin{aligned} -x_1 + m_{11} z_1 &= f_1, \\ -x_2 + m_{21} z_1 &= f_2, \\ -x_3 + m_{31} z_1 &= f_3, \end{aligned} \tag{5.31}$$

which could be reduced to

$$\begin{aligned} \frac{m_{21}}{m_{11}} (f_1 + x_1) &= (f_2 + x_2), \\ \frac{m_{31}}{m_{11}} (f_1 + x_1) &= (f_3 + x_3). \end{aligned} \tag{5.32}$$

Two linear equations in 3 unknowns $(f_1 + x_1)$, $(f_2 + x_2)$, and $(f_3 + x_3)$ is the expression of a straight line in 3-dimensional space. In general, the point (f_1, f_2, f_3) defined by the observations will not lie on this line. The desired correction (x_1, x_2, x_3) is obviously the one which is the shortest distance from the point to the line, since this implies the minimum correction to the observations. If the coordinate axes are rectangular and the scale is the same in all directions, then this minimum will be

$$\sum_i x_i^2 = \text{Min.} \tag{5.33}$$

However, the equations (5.27) can just as well refer to oblique axes with a different scale along each coordinate axis, in which case we must rewrite (5.33) as

$$\sum_{i\,j} g_{ij} x_i x_j = \text{Min} \tag{5.34}$$

or, in matrix form,

$$\mathbf{x}^T\mathbf{G}\mathbf{x} = \text{Min.} \tag{5.35}$$

The next problem is to determine, given a set of observations O_i with associated variances σ_i^2 and covariances $\text{Cov}\,\{O_i, O_j\}$, the appropriate quantities g_{ij} to use in (5.34), (5.35). If the observations are uncorrelated—that is, all covariances zero—then the obvious choice is

$$\begin{aligned} g_{ii} &= 1/\sigma_i^2, \\ g_{ij} &= 0, \qquad i \neq j, \end{aligned} \tag{5.36}$$

since the square of the correction applied would thus be weighted so as to be proportionate to its expected mean square magnitude. If we have a set of n correlated observations $\mathbf{x}$, we can always find a linear transformation thereof [see (2.3)] to another set of n observations $\mathbf{y}$ that are uncorrelated, because in order to determine the n^2 numbers in the transformation matrix $\mathbf{A}$,

$$\mathbf{y} = \mathbf{A}\mathbf{x}, \tag{5.37}$$

there are n^2 condition equations to be satisfied in the relationship between the covariance matrices of $\mathbf{y}$ and $\mathbf{x}$, $\mathbf{W}_y$ and $\mathbf{W}_x$:

$$\mathbf{W}_y = \mathbf{A}\mathbf{W}_x\mathbf{A}^T; \tag{5.38}$$

that is,

$$\begin{aligned} \sigma^2\{y_i\} &= \sum_{k,l} a_{ik}\,\text{Cov}\,\{x_k, x_l\}a_{il}, \\ 0 &= \sum_{k,l} a_{ik}\,\text{Cov}\,\{x_k, x_l\}a_{jl}, \end{aligned} \qquad i \neq j. \tag{5.39}$$

Then the inverse variance weighting (5.36) applied to the quadratic sum minimization of $\mathbf{y}$ implies

$$\begin{aligned} \text{Min} &= \mathbf{y}^T\mathbf{W}_y^{-1}\mathbf{y} = (\mathbf{A}\mathbf{x})^T[\mathbf{A}\mathbf{W}_x\mathbf{A}^T]^{-1}(\mathbf{A}\mathbf{x}) \\ &= \mathbf{x}^T\mathbf{A}^T(\mathbf{A}^T)^{-1}\mathbf{W}_x^{-1}\mathbf{A}^{-1}\mathbf{A}\mathbf{x} \\ &= \mathbf{x}^T\mathbf{W}_x^{-1}\mathbf{x} = \text{Min.} \end{aligned} \tag{5.40}$$

The weight matrix $\mathbf{G}$ used in (5.35) should thus be the inverse of the covariance matrix $\mathbf{W}$.

As stated, the matrix form of the observation equations, (5.27), appears adequate for the problems in geodetic use of satellites. However, for other possible uses it is instructive to solve, subject to (5.40), a more general case,

$$\mathbf{C}\mathbf{x} + \mathbf{M}\mathbf{z} = \mathbf{f}, \tag{5.41}$$

in which there is at least one nonzero element per row in $\mathbf{C}$. Combining (5.40) and (5.41), we can write

$$\mathbf{x}^T\mathbf{W}^{-1}\mathbf{x} - 2(\mathbf{Cx} + \mathbf{Mz} - \mathbf{f})^T\boldsymbol{\lambda} = \text{Min}, \tag{5.42}$$

where $\boldsymbol{\lambda}$ is a vector of parameters called Lagrangian multipliers. We differentiate (5.42) with respect to $\mathbf{x}$, and then set the result equal to zero in order to obtain the minimum

$$\mathbf{W}^{-1}\mathbf{x} - \mathbf{C}^T\boldsymbol{\lambda} = \mathbf{0}. \tag{5.43}$$

If we now solve (5.43) for $\mathbf{x}$ and substitute the result in (5.41), we get

$$\mathbf{K}\boldsymbol{\lambda} + \mathbf{Mz} - \mathbf{f} = \mathbf{0}, \tag{5.44}$$

where

$$\mathbf{K} = \mathbf{CWC}^T. \tag{5.45}$$

On differentiating (5.42) with respect to $\mathbf{z}$, and setting the result equal to zero,

$$\mathbf{M}^T\boldsymbol{\lambda} = \mathbf{0}. \tag{5.46}$$

If we solve (5.44) for $\boldsymbol{\lambda}$,

$$\boldsymbol{\lambda} = \mathbf{K}^{-1}(\mathbf{f} - \mathbf{Mz}). \tag{5.47}$$

By substituting from (5.47) for $\boldsymbol{\lambda}$ in (5.46) and solving for $\mathbf{z}$,

$$\mathbf{z} = [\mathbf{M}^T\mathbf{K}^{-1}\mathbf{M}]^{-1}\mathbf{M}^T\mathbf{K}^{-1}\mathbf{f}. \tag{5.48}$$

The matrix $\mathbf{M}^T\mathbf{K}^{-1}\mathbf{M}$ is called the normal equation coefficients, and the vector $\mathbf{M}^T\mathbf{K}^{-1}\mathbf{f}$, the normal equation constants.

By substituting from (5.48) for $\mathbf{z}$ in (5.47), solving (5.43) for $\mathbf{x}$, and then substituting from (5.47) for $\boldsymbol{\lambda}$ in the result, we have

$$\mathbf{x} = \mathbf{WC}^T\mathbf{K}^{-1}[\mathbf{I} - \mathbf{M}(\mathbf{M}^T\mathbf{K}^{-1}\mathbf{M})^{-1}\mathbf{M}^T\mathbf{K}^{-1}]\mathbf{f}. \tag{5.49}$$

Given $\mathbf{W}$ is the covariance matrix of the observations before correction by $\mathbf{x}$, we find that the covariance matrix of the residuals $\mathbf{f}$ will be, from (5.41), $\mathbf{CWC}^T$, $\mathbf{K}$, by (5.45). To obtain the covariance matrix $\mathbf{V}$ of $\mathbf{z}$, we pre- and post-multiply $\mathbf{K}$ by the coefficient of $\mathbf{f}$ in (5.48):

$$\begin{aligned}\mathbf{V}_z &= [(\mathbf{M}^T\mathbf{K}^{-1}\mathbf{M})^{-1}\mathbf{M}^T\mathbf{K}^{-1}]\mathbf{K}[(\mathbf{M}^T\mathbf{K}^{-1}\mathbf{M})^{-1}\mathbf{M}^T\mathbf{K}^{-1}]^T \\ &= (\mathbf{M}^T\mathbf{K}^{-1}\mathbf{M})^{-1}.\end{aligned} \tag{5.50}$$

Similarly, for the covariance matrix $\mathbf{U}$ of the corrections $\mathbf{x}$, pre- and post-multiply $\mathbf{K}$ by the coefficient of $\mathbf{f}$ in (5.49). After sorting out the algebra, we have

$$\mathbf{U} = \mathbf{WC}^T\mathbf{K}^{-1}[\mathbf{I} - \mathbf{M}(\mathbf{M}^T\mathbf{K}^{-1}\mathbf{M})^{-1}\mathbf{M}^T\mathbf{K}^{-1}]\mathbf{CW}. \tag{5.51}$$

The covariance matrix $\mathbf{V}_x$ of the improved observations then will be

$$\mathbf{V}_x = \mathbf{W} - \mathbf{U}. \tag{5.52}$$

All the results (5.48)–(5.52) are applicable to (5.27) by setting $-\mathbf{I}$ in place of $\mathbf{C}$.

The only assumption we have made in Equations (5.33)–(5.52) is that in the uncorrelated case the weighting factor is inversely proportional to the variance. Quadratic sum minimization under this assumption is known as *minimum variance* and is less restrictive than *maximum likelihood*, where the same result requires assuming that the corrections have a normal distribution about a zero mean: that is, a frequency for a correction of magnitude x proportionate to $\exp(-x^2/\sigma^2)$. However, any improvement over this normal law would require higher degree terms than quadratic in the minimized sum (5.34) or (5.35), which in turn would imply that nonlinear effects from sources unaccounted for by the model are significant. This implication in a way contradicts the assumption that nonlinear Taylor series terms in the model effects (5.29) are negligible. While there are physical situations where there are limitations on $(O_i + dO_i)$, such as that it cannot be negative, which might make a skew distribution of dO_i appropriate, usually a non-normal distribution of corrections dO_i is an indicator of some condition that can be removed in a determinate manner by improving the model rather than by complicating the statistics.

The normal equation coefficient matrix $\mathbf{M}^T\mathbf{K}^{-1}\mathbf{M}$ that is inverted in the solution (5.48) for corrections to parameters $\mathbf{z}$ may be of considerable dimension because of the large number of independent parameters that have perceptible effects on satellite orbits, and because it may be desirable to combine several different orbits in order to get a well-conditioned solution for parameters common to all orbits, such as the gravitational field coefficients and station position shifts. However, if the normal equation coefficient matrix $\mathbf{M}^T\mathbf{K}^{-1}\mathbf{M}$ has a particular form, an appreciable reduction can be obtained in the size of the matrix which must be stored and inverted by the computer. Let

$$\mathbf{N} = \mathbf{M}^T\mathbf{K}^{-1}\mathbf{M}, \qquad \mathbf{s} = \mathbf{M}^T\mathbf{K}^{-1}\mathbf{f}, \tag{5.53}$$

and let

$$\begin{bmatrix} \mathbf{z}_1 \\ \hdashline \mathbf{z}_2 \end{bmatrix} = \mathbf{z}, \quad \begin{bmatrix} \mathbf{s}_1 \\ \hdashline \mathbf{s}_2 \end{bmatrix} = \mathbf{s}, \quad \left[\begin{array}{c|c} \mathbf{N}_{11} & \mathbf{N}_{12} \\ \hline \mathbf{N}_{21} & \mathbf{N}_{22} \end{array}\right] = \mathbf{N}. \tag{5.54}$$

Equation (5.48) can now be written

$$\begin{aligned} \mathbf{N}_{11}\mathbf{z}_1 + \mathbf{N}_{12}\mathbf{z}_2 &= \mathbf{s}_1, \\ \mathbf{N}_{21}\mathbf{z}_1 + \mathbf{N}_{22}\mathbf{z}_2 &= \mathbf{s}_2. \end{aligned} \tag{5.55}$$

Solving the second equation of (5.55) for $\mathbf{z}_2$, we get

$$\mathbf{z}_2 = \mathbf{N}_{22}^{-1}(\mathbf{s}_2 - \mathbf{N}_{21}\mathbf{z}_1), \tag{5.56}$$

and substituting the result in the first equation of (5.55), we have

$$(\mathbf{N}_{11} - \mathbf{N}_{12}\mathbf{N}_{22}^{-1}\mathbf{N}_{21})\mathbf{z}_1 = \mathbf{s}_1 - \mathbf{N}_{12}\mathbf{N}_{22}^{-1}\mathbf{s}_2. \tag{5.57}$$

If $\mathbf{N}$ has the form

$$\mathbf{N} = \left[\begin{array}{c|cccc} \mathbf{N}_{11} & \mathbf{N}_{12,1} & \mathbf{N}_{12,2} & \cdots & \mathbf{N}_{12,n} \\ \hline \mathbf{N}_{21,1} & \mathbf{N}_{22,11} & \mathbf{0} & \cdots & \mathbf{0} \\ \mathbf{N}_{21,2} & \mathbf{0} & \mathbf{N}_{22,22} & & \vdots \\ \vdots & \vdots & & \ddots & \\ \mathbf{N}_{21,n} & \mathbf{0} & \cdots & & \mathbf{N}_{22,nn} \end{array}\right], \tag{5.58}$$

namely,

$$\mathbf{N}_{22,ij} = \mathbf{0}, \qquad i \neq j, \tag{5.59}$$

then the terms involving inversions in (5.57) can be written

$$\begin{aligned} \mathbf{N}_{12}\mathbf{N}_{22}^{-1}\mathbf{N}_{21} &= \sum_{i=1}^{n} \mathbf{N}_{12,i}\mathbf{N}_{22,ii}^{-1}\mathbf{N}_{21,i}, \\ \mathbf{N}_{12}\mathbf{N}_{22}^{-1}\mathbf{s}_2 &= \sum_{i=1}^{n} \mathbf{N}_{12,i}\mathbf{N}_{22,ii}^{-1}\mathbf{s}_{2,i}. \end{aligned} \tag{5.60}$$

If in the matrix of observation Equations (5.27) $\mathbf{M}$ has the form

$$\mathbf{M} = \left[\begin{array}{c|cccc} \mathbf{M}_{11} & \mathbf{M}_{12,1} & 0 & \cdots & 0 \\ \mathbf{M}_{21} & \mathbf{0} & \mathbf{M}_{12,2} & & \vdots \\ \vdots & \vdots & & \ddots & \\ \mathbf{M}_{l1} & \mathbf{0} & \cdots & & \mathbf{M}_{12,l} \end{array}\right], \tag{5.61}$$

$\mathbf{N}$ will have the form (5.58) since, from (5.53),

$$\mathbf{N}_{22,ij} = \sum_{h=1}^{l} \mathbf{M}_{h2,i}^{T}\mathbf{K}_{hh}^{-1}\mathbf{M}_{h2,j}, \tag{5.62}$$

which is $\mathbf{0}$ for $i \neq j$. The matrix $\mathbf{M}$ will have the form (5.61) if several different satellite orbits, each with its own constants of integration, are combined in a single least squares solution, because the set of observation equations peculiar to a particular orbital arc h will contain partial derivatives $\mathbf{M}_{h1}$ with respect to gravitational coefficients, and so on, common to all orbits, but have nonzero partial derivatives with respect to only its own orbital constants, $\mathbf{M}_{h2,h}$. Hence, a solution for n parameters common to any number of orbits each with p constants of integration can be made in accordance with (5.48) without storing any matrix of dimension larger than $(n + p)$ or inverting any matrix of dimension larger than n.

The vectors $\mathbf{x}$ and $\mathbf{z}$ may be corrections to "observations" or "parameters" that occur in a staged or evolutionary process. If the vectors of the actual errors at the ith stage are $\boldsymbol{\epsilon}_i(\mathbf{x})$, $\boldsymbol{\epsilon}_i(\mathbf{z})$, and their nonlinear effects at the $(i + 1)$th stage are insignificant, then the errors $\boldsymbol{\epsilon}_{i+1}(\mathbf{x})$, $\boldsymbol{\epsilon}_{i+1}(\mathbf{z})$ can be expressed as a linear transform of $\boldsymbol{\epsilon}_i(\mathbf{x})$, $\boldsymbol{\epsilon}_i(\mathbf{z})$ through propagation matrices $\mathbf{P}_x$, $\mathbf{P}_z$:

$$\begin{aligned} \boldsymbol{\epsilon}_{i+1}(\mathbf{x}) &= \mathbf{P}_x\boldsymbol{\epsilon}_i(\mathbf{x}), \\ \boldsymbol{\epsilon}_{i+1}(\mathbf{z}) &= \mathbf{P}_z\boldsymbol{\epsilon}_i(\mathbf{z}). \end{aligned} \tag{5.63}$$

Consequently the covariance matrices $\mathbf{W}_x$, $\mathbf{W}_z$ at the $(i + 1)$th stage can be expressed as

$$\begin{aligned} \mathbf{W}_x &= \mathbf{P}_x\mathbf{V}_x\mathbf{P}_x^T, \\ \mathbf{W}_z &= \mathbf{P}_z\mathbf{V}_z\mathbf{P}_z^T. \end{aligned} \tag{5.64}$$

If these variables are to be further corrected at the $(i + 1)$th stage by being combined with new observations with covariance matrix $\mathbf{W}_y$ in new condition equations, the equations can be written as

$$\mathbf{C}_x\mathbf{x}_x + \mathbf{C}_y\mathbf{x}_y + \mathbf{C}_z\mathbf{x}_z + \mathbf{M}\mathbf{z} = \mathbf{f} \tag{5.65}$$

and the quadratic sum to be minimized as

$$\mathbf{x}_x^T\mathbf{W}_x^{-1}\mathbf{x}_x + \mathbf{x}_y^T\mathbf{W}_y^{-1}\mathbf{x}_y + \mathbf{x}_z^T\mathbf{W}_z^{-1}\mathbf{x}_z = \text{Min}. \tag{5.66}$$

Particular cases to which (5.64)–(5.66) might be applied are:

1. An evolutionary process, such as an orbit, in which the carried-forward estimate of the state of the process, with covariance matrix $\mathbf{W}_x$, is combined with new observations with covariance matrix $\mathbf{W}_y$. In this case, $\mathbf{C}_z$ is $\mathbf{0}$, $\mathbf{M}$ is $\mathbf{0}$, and the solution by (5.49) becomes

$$\begin{bmatrix} \mathbf{x}_x \\ \hline \mathbf{x}_y \end{bmatrix} = \begin{bmatrix} \mathbf{W}_x\mathbf{C}_x^T \\ \hline \mathbf{W}_y\mathbf{C}_y^T \end{bmatrix} \{\mathbf{C}_x\mathbf{W}_x\mathbf{C}_x^T + \mathbf{C}_y\mathbf{W}_y\mathbf{C}_y^T\}^{-1}\mathbf{f}. \tag{5.67}$$

2. Corrections to parameters may be determined from several sets of observations each of which is ill-conditioned alone, but which are inconvenient to combine. In order to express the effect of the corrections and associated covariance matrix $\mathbf{V}_z$ from one set in analyzing another set, we can put $\mathbf{W}_x$ and $\mathbf{C}_x$ as $\mathbf{0}$; $\mathbf{P}_z$ as $\mathbf{I}$, and $\mathbf{C}_y$, $\mathbf{C}_z$, $\mathbf{M}$, and $\mathbf{f}$ as

$$\mathbf{C}_y = \begin{bmatrix} \mathbf{C}_y \\ \hdashline \mathbf{0} \end{bmatrix}, \quad \mathbf{C}_z = \begin{bmatrix} \mathbf{0} \\ \hdashline -\mathbf{I} \end{bmatrix}, \quad \mathbf{M} = \begin{bmatrix} \mathbf{M}_y \\ \hdashline \mathbf{I} \end{bmatrix}, \quad \mathbf{f} = \begin{bmatrix} \mathbf{f}_y \\ \hdashline \mathbf{0} \end{bmatrix}. \tag{5.68}$$

The solution by (5.48) becomes

$$\mathbf{z} = [\mathbf{M}_y^T \mathbf{K}_y^{-1} \mathbf{M}_y + \mathbf{V}_z^{-1}]^{-1} \mathbf{M}_y^T \mathbf{K}_y^{-1} \mathbf{f}_y. \tag{5.69}$$

To summarize, for the geodetic satellite problem the observational data can be transformed or combined so that there is one "observation" per equation as in (5.27); it is the corrections $\mathbf{z}$ to model parameters in which we are interested; and the inadequacies of the model resulting from drag, higher gravitational harmonics, and so on, are such that in an ideal solution there should be appreciable correlation between observations at different times, as expressed by the covariances (5.24), (5.25) that appear as off-diagonal elements in the covariance matrix $\mathbf{W}$.

REFERENCES

1. Apostol, Tom M. *Calculus*. Vol. II. New York: Blaisdell Publishing Company, 1962.

2. Arley, N., and K. R. Buch. *Probability and Statistics*. New York: John Wiley and Sons, 1950.

3. Bartlett, M. S. *An Introduction to Stochastic Processes*. London: Cambridge University Press, 1956.

4. Blackman, R. B. "Methods of Orbit Refinement." *Bell System Tech. J.*, *43* 1964, pp. 885–909.

5. Blackman, R. B., and J. W. Tukey. *The Measurement of Power Spectra*. New York: Dover Publications, 1959.

6. Brown, D. C. "A Treatment of Analytical Photogrammetry with Emphasis on Ballistic Camera Applications." *RCA Data Reduction Tech. Rept. 39*, (1957), 147 pp.

7. Kalman, R. E. "A New Approach to Linear Filtering and Prediction Problems." *Trans. ASME, D, J. Basic Engr.*, *82* (1960) pp. 35–45.

8. Kaula, W. M. "Determination of the Earth's Gravitational Field." *Revs. Geophys.*, *1* (1963), pp. 507–552.

9. Yaglom, A. M. *An Introduction to the Theory of Stationary Random Functions*. Englewood Cliffs, N. J.: Prentice-Hall, Inc., 1962.

6

DATA ANALYSIS

6.1. Simultaneous Observations

If sufficient simultaneous observations are made from the same set of stations, then the corrections to the satellite positions $d\mathbf{x}$ in Equations (4.37) and (4.52) can be regarded as unknowns. Also appearing as unknowns will be the corrections to station positions $d\mathbf{u}_0$ with the exception of one station, since without using the orbit there is no other way of fixing the system with respect to the origin. Hence if there are m stations and n simultaneous sets of observations, there will be $[3(m - 1) + 3n]$ unknowns and qnm observations, where q is 2 for camera observations and 1 for range observations. Hence it is necessary that

$$qnm > 3(m - 1) + 3n \tag{6.1}$$

or

$$n > \frac{3m - 3}{qm - 3}, \tag{6.2}$$

(except for qm 2 or 3). For camera observations, let the $2m$ observations of the jth satellite point be numbered $P_{1j}, P_{2j} \ldots P_{2-j}$; and the satellite coordinates be numbered x_{1j}, x_{2j}, x_{3j}. Then for the matrix set up of the observations (5.27), we have for each simultaneous set from the kth station:

$$\mathbf{M}_{xjk} = \begin{bmatrix} \partial P_{(2k-1)j}/\partial x_{1j}, & \partial P_{(2k-1)j}/\partial x_{2j}, & \partial P_{(2k-1)j}/\partial x_{3j} \\ \partial P_{2kj}/\partial x_{1j}, & \partial P_{2kj}/\partial x_{2j}, & \partial P_{2kj}/\partial x_{3j} \end{bmatrix},$$

$$\mathbf{M}_{ujk} = \begin{bmatrix} \partial P_{(2k-1)j}/\partial u_{1k}, & \partial P_{(2k-1)j}/\partial u_{2k}, & \partial P_{(2k-1)j}/\partial u_{3k} \\ \partial P_{2kj}/\partial u_{1k}, & \partial P_{2kj}/\partial u_{2k}, & \partial P_{2kj}/\partial u_{3k} \end{bmatrix},$$

$$\mathbf{M}_{xj} = \begin{bmatrix} \mathbf{M}_{xj1} \\ \mathbf{M}_{xj2} \\ \cdot \\ \cdot \\ \cdot \\ \mathbf{M}_{xjm} \end{bmatrix}, \quad \mathbf{M}_{uj} = \begin{bmatrix} \mathbf{M}_{uj1} & \mathbf{O} & \cdots & \mathbf{O} \\ \mathbf{O} & \mathbf{M}_{uj2} & & \cdot \\ \cdot & & \cdot & \cdot \\ \cdot & & \cdot & \cdot \\ \cdot & & & \cdot \\ \mathbf{O} & & & \mathbf{M}_{uj(m-1)} \\ \mathbf{O} & & \cdots & \mathbf{O} \end{bmatrix}, \tag{6.3}$$

$$\mathbf{M} = \left[\begin{array}{cccc|c} \mathbf{M}_{x1} & \mathbf{O} & \cdots & \mathbf{O} & \mathbf{M}_{u1} \\ \mathbf{O} & \mathbf{M}_{x2} & & & \mathbf{M}_{u2} \\ \cdot & & \cdot & & \cdot \\ \cdot & & & \cdot & \cdot \\ \cdot & & & & \cdot \\ \mathbf{O} & & \cdots & \mathbf{M}_{xn} & \mathbf{M}_{un} \end{array}\right], \quad \mathbf{f} = \begin{bmatrix} \mathbf{f}_1 \\ \mathbf{f}_2 \\ \cdot \\ \cdot \\ \cdot \\ \mathbf{f}_n \end{bmatrix};$$

$\mathbf{M}_{xjk}$ is the coefficient

$$\begin{bmatrix} f/p_3 & 0 & 0 \\ 0 & f/p_3 & 0 \end{bmatrix} \mathbf{R}_{px}$$

of $d\mathbf{x}$ in (4.37), and $\mathbf{M}_{ujk}$ is the coefficient

$$-\begin{bmatrix} f/p_3 & 0 & 0 \\ 0 & f/p_3 & 0 \end{bmatrix} \mathbf{R}_{px} R_3(-\theta)$$

of $d\mathbf{u}_0$ in (4.37).

If the observations are considered to be uncorrelated, then the matrix $\mathbf{W}^{-1}$ (the same as $\mathbf{K}^{-1}$ if $\mathbf{C} = \mathbf{I}$) in (5.48) will be diagonal. Letting $\boldsymbol{\sigma}$ be the 2×2 diagonal matrix of variances for each photograph, then in the quantities $\mathbf{M}^T\mathbf{W}^{-1}\mathbf{M}$ and $\mathbf{M}^T\mathbf{W}^{-1}\mathbf{f}$ can be incremented at each observation by

$$\mathbf{M}_{jk}^T \boldsymbol{\sigma}^{-1} \mathbf{M}_{jk} \quad \text{and} \quad \mathbf{M}_{jk}^T \boldsymbol{\sigma}^{-1} \mathbf{f}_{jk},$$

where

$$\mathbf{M}_{jk} = [0 \cdots \mathbf{M}_{xjk} \cdots 0 \mid 0 \cdots \mathbf{M}_{ujk} \cdots 0].$$

Since no reference is made to the center of mass of the earth, the $\mathbf{x}$, $\mathbf{u}$ coordinate systems need not be used; instead, some locally referred system, such as the $\mathbf{l}$ coordinates of the fixed station, can be used, applying rotations as described by (4.17).

Simultaneous observations have been made by fixed cameras of both balloon-type and flashing light satellites.

6.2. Orbital Observations: Short-Term

If observations are nonsimultaneous, then, of course, the orbital constants of integration must be added to the parameters in place of satellite positions. In addition, there may be other orbital parameters, including those of particular interest to geodesy:the coefficients of the gravitational field. All station positions may be considered free, since the absence of first-degree harmonics from the gravitational field is equivalent to assuming the geometrical center of the earth, to which station positions refer, coincides with the dynamical center of mass, to which the orbit refers.

Hence, ideally, in (5.27), the **M** matrix should now include partial derivatives of the calculated quantities C_i with respect to the orbital parameters and the gravitational coefficients as obtained by combining (4.29) and (4.37) or (4.48), and the part C_i of the **f** vector should be calculated from the orbital theory by (3.76) and (3.113) plus the appropriate transformation to observational form: (4.15) or (4.36). In (5.48), the covariance matrix **W** should now ideally include off diagonal elements expressing covariance, as well as increments to the variances on the main diagonal, due to the effects of neglected higher gravitational coefficients by (5.24) and of drag by (5.26).

The accuracy of the principal types of observation is such that it should be hoped to determine orbital oscillations on the order of ± 10 meters. For typical orbit specifications, the order of magnitude of spherical harmonics suggested by (5.14) used in (3.76) indicates that tesseral harmonics ($m \neq 0$) as high as the 8th degree and most up to the 6th degree will be determinable, namely, up to 60 coefficients. However, the accuracy of location of tracking stations with respect to each other is poor enough that the ± 10 meters criterion indicates that their coordinates should also be treated as unknown, namely, about 36 more parameters. Furthermore, the principal effects of some sets of terms with the same order subscript m are all exactly in phase, since they all have the same argument, $m(\Omega - \theta)$, in the partial derivative (4.27): for example, (l, m) of (2, 2), (4, 2), (6, 2), and (8, 2). To distinguish such terms from each other, the coefficients K_{ilmpq} of (4.28) must differ. Varying the semimajor axis a, or eccentricity e, appreciably will vary the drag characteristics by a considerable amount; hence, it is desirable to obtain the variety of orbital specifications to separate terms of the same argument by varying the inclination. We thus must add to the total parameters the elements of enough orbits to make this separation—say four orbits, or 28 elements, including in each set a parameter to absorb acceleration in the mean anomaly.

For a fairly firm determination of this total of more than 120 parameters, several hundred observations are required. The covariance matrix **W** of these observations will be a series of nonzero blocks down the principal diagonal, one block for each orbit, with off-diagonal elements due to drag effect according to (5.26). This rigorous treatment has not yet been applied; in practice, the covariance matrix **W** of such large dimensions with nonzero off-diagonal elements due to drag effect according to (5.26) is beyond reasonable computer capacity. Hence, in practice, the rigorous treatment has not yet been applied; the covariance matrix **W** is either taken as a diagonal matrix, or, at best, covariance is taken into account only between observations in the same pass. Furthermore, the lack of a sufficient variety in inclination of orbits at a good altitude has degraded the accuracy to be reasonably hoped for to about ± 20 meters, which reduces the number of low degree gravitational coefficients accurately determinable to about 35: all up to an (l, m) of (4, 4); plus (5, 0), (5, 1); (6, 0) through (6, 4); and (7, 0). In addition to these low degree coefficients, one pair of coefficients in the range of about (9, 9) to (15, 15) should be added for each satellite to absorb the small-divisor effect described by (3.150).

The various aforestated difficulties have the result that progress in orbit analysis for tesseral harmonics and station position has been by computer experimentation: the testing out in actual computation of various alternative procedures selected by empirical rules. Some of these procedures and rules are given as follows:

1. The length of the orbital arc represented by a single set of constants of integration may be selected either on the basis that there is a considerable surplus of observations over parameters to be determined—for example, 80 or more observations—or that the orbital residuals are not more than a small multiple of the amplitudes of the perturbations caused by the tesseral harmonics—for example, 10 to 1. Rules of this sort generally result in arcs for Doppler tracking, which can observe in all conditions, of one to seven days, but for camera tracking, which requires a combination of satellite in sun and station in darkness and clear weather, of ten to thirty days.

2. The residuals can usually be reduced by using arbitrary polynomials in time to represent some of the variation of the elements. However, to avoid absorbing some of the effects of the gravitational variations, such polynomials are usually confined to a t^2, or a t^2 and a t^3 term in the mean anomaly, where the maximum drag effect occurs.

3. To further minimize drag effect, the across track component may be given higher weight (or lower variance) than the along track component. This may most conveniently be done using the partial derivative of the observation with respect to time, (4.34). For example, for camera observations with components p_1, p_2 we have

$$V_{\text{Obs}} = \begin{bmatrix} \sigma_p^2 & 0 \\ 0 & \sigma_p^2 \end{bmatrix} + \begin{bmatrix} \partial p_1/\partial \epsilon(t) \\ \partial p_2/\partial \epsilon(t) \end{bmatrix} \sigma^2(t) \left[\frac{\partial p_1}{\partial \epsilon(t)}, \quad \frac{\partial p_2}{\partial \epsilon(t)} \right]. \tag{6.4}$$

Along-track residuals generally are about twice as large as across-track; however, some experience has indicated that weighting such as (6.4) does not make much difference in results.

4. Since nonuniform distribution of observations destroys the orthogonality of the low-frequency drag variations in Figure 16 to the higher frequency gravitational variations, presumably some of the separation can be restored by weighting observations inversely as their density with respect to phase angles important in determining the gravitational coefficients, such as $(\Omega - \theta)$. Also, if tracking stations are nonuniformly distributed geographically, observations that are from those stations clustered closely together may be accorded lower weight than those from remote stations. Again, the device of weighting to overcome nonuniform distribution is one whose benefits have been difficult to discern in application.

5. The number and distribution of observations will often be such as to cause ill-conditioning: that is, the effects of different parameters are so similar as to make them difficult to distinguish. One technique to reduce ill-conditioning is to use a preassigned covariance matrix for the parameters, as in (5.69), using statistical estimates such as (5.14). This technique probably tends to reduce the magnitude of the results below that of the actual values.

The ideal method of removing ill-conditioning is to include data from orbits of several different inclinations. In order to avoid having too large matrices to invert, the calculation has sometimes been done by the partitioned solution of the normals described by (5.53)–(5.62), and sometimes in two separate steps. In the first step, the orbital constants of integration are determined separately for each arc. Subtracting out the effects of these constants leaves residuals which are then analyzed jointly for all orbits to determine the gravitational coefficients and station coordinate shifts. This technique probably tends to cause some effects of the gravitational variations to be absorbed at the earlier step by the orbital constants of integration.

6. Tests that can be applied in the analysis include (a) the solution for the gravitational coefficients C_{21}, S_{21} that are known to be virtually zero from the smallness of the variations in the position of the earth's rotation axis; and (b) comparing the geoid height calculated geometrically from the station coordinates—that is, determining the altitude h in (4.5) from given u, v, w and subtracting the height above sea level therefrom—with that determined from the gravitational coefficients. Thus

$$N = R_E \sum_{l,m} P_{lm}(\cos \phi)[C_{lm} \cos m\lambda + S_{lm} \sin m\lambda]. \tag{6.5}$$

The results that have been obtained thus far from satellite orbits for the variations of the gravity field have plainly been influenced by the method of analysis as much as by the orbital characteristics and the type of tracking.

In general, however, results obtained from Doppler tracking yield larger mean square coefficients than those from optical tracking. Perhaps this happens because there is more systematic observational error in the Doppler tracking; or perhaps because the much scantier and less uniform distribution of camera observations results in a greater part of the effects of the variations of the gravitational field being absorbed by the orbital constants of integration; or perhaps because the Doppler data included more high inclination orbits, which are more sensitive to the tesseral harmonics, particularly the sectoral ($m = l$) terms. Investigations of statistical interaction between different parameters show that the highest correlation is, as expected, between coefficients of subscript (l, m), (n, m), $l - n$ even; that the correlation is rather low between station coordinates and gravitational coefficients, or between different station coordinates; and that there is a moderate amount of correlation between gravitational coefficients and orbital constants of integration. The present rate of improvement in results is rapid, partly because of more tracking data under better conditions and partly because of better methods of analysis. The results given in Table 3 were all obtained in the fall of 1964. Figure 17 is the geoid computed by using the coefficients of Anderle (1966) from Table 3, plus zonal harmonics $\bar{C}_{30}$ through $\bar{C}_{60}$, in (6.5). All solutions since 1963 show a strong resemblance in the locations of the principal extrema of the gravitational field. There are always four maxima: (1) near New Guinea (0°, 150°E), (2) near Great Britain (50°N, 10°W), (3) off the Cape of Good Hope (50°S, 40°E), and (4) near Peru (10°S, 80°W); and four or five minima: (1) off India (0°, 70°E), (2) near the South Pole (90°S), (3) in the western Atlantic (20°N, 60°W), (4) in the eastern Pacific (20°N, 120°W), and sometimes (5) in the northwestern Pacific (40°N, 180°).

A compromise between simultaneous observations and one-to-thirty day arcs which has been occasionally proposed is an intensively observed short arc of about 30 minutes duration to relate the positions of two geodetic datums. Although the minimum of six orbital elements must still be included as unknown parameters, at least one datum must be considered fixed because the arc is too short to establish location with respect to the center of mass. The shortness of the arc, the number of observations, and the accuracy requirements make numerical integration of the orbit appropriate. The convenient statistical treatment is then that of the evolutionary process given by (5.64) and (5.67). The advantage of the short arc is that imperfectly known perturbations by gravitational variations, drag, and so on, have had relatively little time to build up. However, these environmental factors are still of sufficient influence that parameters to express their effects must be added to the dynamically necessary six orbital elements as part of the estimate of the state of the process. Since the parameters at one stage do not have a deterministic relationship to those at another stage, the covariance cannot

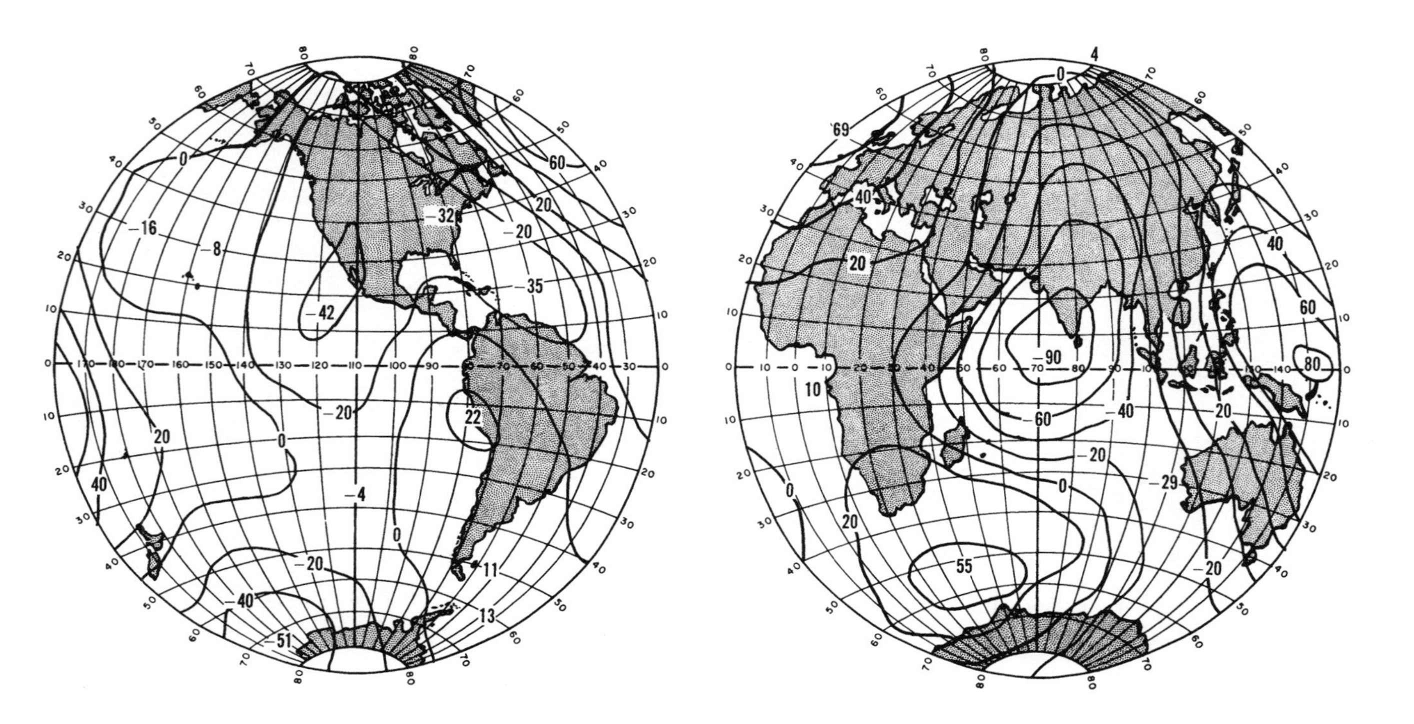

FIGURE 17. *Geoid heights, in meters, based on the spherical harmonic coefficients through the sixth degree of Anderle (1966).*

TABLE 3

Tesseral Harmonic Coefficients of the Gravitational Field

	From Camera Data	From Doppler Data	
	Izsak	*Guier and Newton*	*Anderle*
Coefficient	(1966b)	(1965)	(1966)
$\bar{C}_{22} \times 10^6$	2.08	2.38	2.45
$\bar{S}_{22} \times 10^6$	−1.25	−1.20	−1.52
$\bar{C}_{31} \times 10^6$	1.60	1.84	2.15
$\bar{S}_{31} \times 10^6$	−0.04	0.21	0.27
$\bar{C}_{32} \times 10^6$	0.38	1.22	0.98
$\bar{S}_{32} \times 10^6$	−0.80	−0.68	−0.91
$\bar{C}_{33} \times 10^6$	−0.17	0.66	0.58
$\bar{S}_{33} \times 10^6$	1.40	0.98	1.62
$\bar{C}_{41} \times 10^6$	−0.38	−0.56	−0.49
$\bar{S}_{41} \times 10^6$	−0.40	−0.44	−0.57
$\bar{C}_{42} \times 10^6$	0.20	0.42	0.27
$\bar{S}_{42} \times 10^6$	0.58	0.44	0.67
$\bar{C}_{43} \times 10^6$	0.69	0.84	1.03
$\bar{S}_{43} \times 10^6$	−0.10	0.00	−0.25
$\bar{C}_{44} \times 10^6$	−0.11	−0.21	−0.41
$\bar{S}_{44} \times 10^6$	0.43	0.19	0.34
$\bar{C}_{51} \times 10^6$	−0.14	0.14	0.03
$\bar{S}_{51} \times 10^6$	−0.04	−0.17	−0.12
$\bar{C}_{52} \times 10^6$	0.24	0.27	0.64
$\bar{S}_{52} \times 10^6$	−0.27	−0.34	−0.33
$\bar{C}_{53} \times 10^6$	−0.67	0.09	−0.39
$\bar{S}_{53} \times 10^6$	0.05	0.10	−0.12
$\bar{C}_{54} \times 10^6$	−0.13	−0.49	−0.55
$\bar{S}_{54} \times 10^6$	0.16	−0.26	0.15
$\bar{C}_{55} \times 10^6$	0.08	−0.03	0.21
$\bar{S}_{55} \times 10^6$	−0.41	−0.67	−0.59
$\bar{C}_{61} \times 10^6$	−0.02	0.00	−0.08
$\bar{S}_{61} \times 10^6$	0.12	0.10	0.19
$\bar{C}_{62} \times 10^6$	0.05	−0.16	0.13
$\bar{S}_{62} \times 10^6$	−0.23	−0.16	−0.46
$\bar{C}_{63} \times 10^6$	0.05	0.53	−0.02
$\bar{S}_{63} \times 10^6$	0.00	0.05	−0.13
$\bar{C}_{64} \times 10^6$	0.07	−0.31	−0.19
$\bar{S}_{64} \times 10^6$	−0.39	−0.51	−0.32
$\bar{C}_{65} \times 10^6$	−0.28	−0.18	−0.09
$\bar{S}_{65} \times 10^6$	−0.38	−0.50	−0.79
$\bar{C}_{66} \times 10^6$	−0.12	0.01	−0.32
$\bar{S}_{66} \times 10^6$	−0.59	−0.23	−0.36

be simply propagated as in (5.64), but must be supplemented by a statistical increment expressing the uncorrelated part of the variance at the two stages.

6.3 Orbital Observations: Long-Term

If the orbital elements have secular change (that is, proportionate to t or t^2), or have sinusoidal variation of much larger period than the arc lengths, then if these effects are not calculated in the determination of an orbit and other parameters as described in Section 6.2 the orbital constants of integration will reflect the long-term secular and periodic changes. A differential correction determination of orbital elements for an arc of a week or two is, in fact, a very effective means of smoothing or filtering short period variations, so that the long-term changes stand out in the constants of integration. As is shown by the linear perturbation formula (3.76), the zonal harmonics C_{l0} of the gravitational field will give rise either to purely secular effects for l even, or to long period effects of argument ω for l odd. Hence determination of these zonal coefficients C_{l0} is best done by analyzing the change in the mean orbital elements of short arcs over several months.

Determination of the even zonal harmonics is made from the motion of the node and sometimes from the motion of the perigee. From (3.76) and (3.113) we have

$$\dot{\Omega} = -\sum_{p=1}^{\infty} J_{2p}\,\mu a_e^{2p} y\,\frac{(\partial F_{2p0p}/\partial i) G_{2pp0}(e)}{na^{2p+3}(1-e^2)^{1/2}\sin i} + O(J_2^2) + \text{lunisolar terms}, \quad (6.6)$$

$$\dot{\omega} = -\sum_{p=1}^{\infty} J_{2p}\,\mu a_e^{2p} y\,\frac{(1-e^2)^{1/2}}{na^{2p+3}}\left[\frac{\partial G_{2pp0}/\partial e}{e} F_{2p0p}(i) - \frac{\cot i}{(1-e^2)}\frac{\partial F_{2p0p}}{\partial i} G_{2pp0}(e)\right] + O(J_2^2) + \text{lunisolar terms}. \quad (6.7)$$

Precautions that have to be observed in the analysis are

1. The set of satellite orbits used should have a variety of inclinations sufficient to separate the different harmonics.
2. The orbital constants of integration determined by differential correction for the short arcs must be consistent with the algebraic form of the terms containing J_2^2.
3. The mean value of the elements a, e, i for the entire duration used to determine $\dot{\Omega}$, $\dot{\omega}$ must be accurately determined, to be valid for use in (6.6) and (6.7). Correct averaging of the elements a and e is particularly important to remove secular drag effects.
4. If lunisolar attraction, radiation pressure, and other perturbations are not removed in determining the mean values of the constants of integration, they can distort determination of the rates $\dot{\Omega}$ and $\dot{\omega}$ not only through purely secular effects but also through periodic perturbations. A periodic

perturbation $\Delta(\Omega, \omega) \sin\{\varkappa t - \lambda\}$ will affect the apparent secular rate from observations lasting from t_1 to t_2 by an amount $\delta(\dot{\Omega}, \dot{\omega})$,

$$\delta(\dot{\Omega}, \dot{\omega}) = \frac{\Delta(\Omega, \omega)[\sin\{\varkappa t_2 - \lambda\} - \sin\{\varkappa t_1 - \lambda\}]}{t_2 - t_1}. \tag{6.8}$$

5. If the perturbations are removed in determining the constants of integration, in addition to direct effects $\Delta_1(\Omega, \omega)$, the interaction of perturbation Δe, Δi with the secular effect of J_2 may cause an indirect effect $\Delta_2(\Omega, \omega)$ large enough that it should be taken into account, as given by (3.116).

TABLE 4

Zonal Harmonic Coefficients of the Gravitational Field

Coefficient	*Smith* (1963, 1965)	*Kozai* (1964)	*King-Hele et al.* (1965ab)	*Guier & Newton* (1965ab)
$J_2 \times 10^6$	1082.64	1082.65	1082.64	
$J_3 \times 10^6$	−2.44	−2.55	−2.56	−2.68
$J_4 \times 10^6$	−1.70	−1.65	−1.52	
$J_5 \times 10^6$	−0.18	−0.21	−0.15	−0.02
$J_6 \times 10^6$	0.73	0.65	0.57	
$J_7 \times 10^6$	−0.30	−0.33	−0.44	−0.59
$J_8 \times 10^6$	−0.46	−0.27	0.44	
$J_9 \times 10^6$		−0.05	0.12	0.18
$J_{10} \times 10^6$	−0.17	−0.05		
$J_{11} \times 10^6$		0.30		
$J_{12} \times 10^6$	−0.22	−0.36		
$J_{13} \times 10^6$		−0.11		
$J_{14} \times 10^6$	0.19	0.18		

Current analyses of secular motions use several months of data each of seven or more satellites. The principal differences in treatment are in the relative weighting: whether according to accuracy or to representation of variety in inclination, and in whether or not perigee motion, which may be more affected by drag, should be used as well as nodal motion. The principal recent determinations are given in Table 4.

The odd degree zonal harmonics are somewhat easier to determine, since more orbital elements are affected and no other orbital perturbations have as argument the perigee angle ω. The most influential perturbation is perigee height. From (3.76) we have

$$\Delta e = -2\sum_{p=2}^{\infty} J_{2p-1} \mu a_e^{2p-1} \frac{F_{(2p-1)0p}(i) G_{(2p-1)01}(e)(1 - e^2)^{1/2}}{n a^{2p+2} \dot{\omega}} \sin\omega. \tag{6.9}$$

Also in the category of long-term orbital observations would be observations of the motion in longitude of a 24-hour satellite, for the purpose of determining $\{C_{22}, S_{22}\}$ through its resonant effect as expressed by (3.136). The length of the expected period T, from (3.137), makes it appropriate to use mean orbital elements determined for epochs on the order of a week apart. From the mean semimajor axis a, eccentricity e, and inclination i, the coefficients Q_{lm} could be calculated by (3.121). Then using these Q_{lm} with observed longitudes λ_A the accelerations $\ddot{\lambda}_A$ could be calculated by (3.126). These calculated $\ddot{\lambda}_A$ would then be compared with the observed $\ddot{\lambda}_A$ obtained by fitting a curve to the observed λ_A.

Wagner (1965) calculated the elements of the first near 24-hour orbit being tracked, SYNCOM II. The mean elements are

$$\begin{aligned} a &= 42{,}170 \text{ km} = 6.61 a_e, \\ e &= 0.0002, \\ i &= 33^\circ. \end{aligned}$$

Calculating the F_{lmp} by Table 1, and taking the G_{lpq} as unity, we get in "planetary" units ($k = 1,\ M = 1,\ a_e = 1$)

$$\begin{aligned} Q_{22} &= 1.206 \times 10^{-3} J_{22} = 0.778 \times 10^{-3} \bar{J}_{22}, \\ Q_{31} &= -0.014 \times 10^{-3} J_{31} = -0.016 \times 10^{-3} \bar{J}_{31}, \\ Q_{33} &= 1.258 \times 10^{-3} J_{33} = 0.175 \times 10^{-3} \bar{J}_{33}, \end{aligned} \tag{6.10}$$

where the $\bar{J}_{lm}$ differ from the J_{lm} by the normalization factor given in (1.34). Satellite SYNCOM II was permitted to drift first from longitude 54.9°W to 57.6°W, then restarted and allowed to drift again from 59.2°W to 63.5°W. Taking the mean longitudes of -56.25° and -61.33° for these two periods, and making the replacement

$$\bar{J}_{lm} \sin m(\lambda - \lambda_{lm}) = \bar{C}_{lm} \sin m\lambda - \bar{S}_{lm} \cos m\lambda, \tag{6.11}$$

we can write expressions for the accelerations in terms of the coefficients and the mean longitudes for each of the two periods,

$$\begin{aligned} \ddot{\lambda}_1 \times 10^3 &= -0.719\bar{C}_{22} + 0.298\bar{S}_{22} + 0.013\bar{C}_{31} \\ &\qquad + 0.009\bar{S}_{31} - 0.034\bar{C}_{33} + 0.172\bar{S}_{33}, \\ \ddot{\lambda}_2 \times 10^3 &= -0.654\bar{C}_{22} + 0.420\bar{S}_{22} + 0.014\bar{C}_{31} \\ &\qquad + 0.007\bar{S}_{31} + 0.012\bar{C}_{33} + 0.175\bar{S}_{33}. \end{aligned} \tag{6.12}$$

Wagner (1965) gives as observed accelerations

$$\begin{aligned}\ddot{\lambda}_{1o} &= -1.27 \pm 0.02 \times 10^{-3} \text{ degrees/day}^2\\ &= -1.93 \pm 0.03 \times 10^{-9} \text{ planetary units,}\\ \ddot{\lambda}_{2o} &= -1.32 \pm 0.02 \times 10^{-3} \text{ degrees/day}^2\\ &= -2.01 \pm 0.03 \times 10^{-9} \text{ planetary units.}\end{aligned}$$

The sets of coefficients from Table 3 used in (6.12) yield:

Coefficients of Izsak (1966): $\ddot{\lambda}_{1c} = -1.59 \times 10^{-9}$ pl units,
$\ddot{\lambda}_{2c} = -1.61 \times 10^{-9}$ pl units,

Coefficients of Guier & Newton (1965): $\ddot{\lambda}_{1c} = -1.86 \times 10^{-9}$ pl units,
$\ddot{\lambda}_{2c} = -1.83 \times 10^{-9}$ pl units,

Coefficients of Anderle (1966): $\ddot{\lambda}_{1c} = -1.92 \times 10^{-9}$ pl units,
$\ddot{\lambda}_{2c} = -1.92 \times 10^{-9}$ pl units.

In addition, there is a small contribution to the acceleration by the sun and moon of about -0.02×10^{-9} pl units.

The approach to resonance expressed by (3.150) was first noted in the orbit of a satellite which had a nodal period of 107.13^m and an inclination of 89.8°. Hence $\dot{\Omega}$ was negligible, and, for $m = 13$,

$$\begin{aligned}\dot{\omega} + \dot{M} = \frac{2\pi \times 806.8137}{107.13 \times 60} &= \quad 0.78866,\\ m(\dot{\Omega} - \dot{\theta}) = -13 \times 0.058834 &= \underline{-0.76484},\\ \dot{\omega} + \dot{M} + m(\dot{\Omega} - \dot{\theta}) &= \quad 0.02382,\end{aligned}$$

or a period of $2\pi/(107.088 \times 0.02382) = 2.47$ days. The eccentricity was 0.003, so the $H(e)\, \partial G_{lpo}/\partial e$ term may be neglected and G_{lpo} set as unity. The semimajor axis was $1.1706a_e$. Evaluating F_{lmp} by (3.62) and setting

$$\bar{S}_{lmpo} = N_{lm}\bar{J}_{lm} \sin [\omega + M + m(\Omega - \theta - \lambda_{lm})], \tag{6.13}$$

where N_{lm} is the normalization factor from (1.34), we get for (3.150)

$$\begin{aligned}\Delta\lambda = &-187.3\, \bar{J}_{13,13} \sin [\omega + M + 13(\Omega - \theta - \lambda_{13,13})]\\ &-79.3\, \bar{J}_{15,13} \sin [\omega + M + 13(\Omega - \theta - \lambda_{15,13})]\\ &+173.2\, \bar{J}_{17,13} \sin [\omega + M + 13(\Omega - \theta - \lambda_{17,13})]\\ &+38.6\, \bar{J}_{19,13} \sin [\omega + M + 13(\Omega - \theta - \lambda_{19,13})]\\ &+ \cdots.\end{aligned} \tag{6.14}$$

From (5.15), the expected order of magnitude of $\bar{J}_{l,m}$ is about $\sqrt{2} \times 10^{-5}/l^2$, and hence 0.07×10^{-6} is a likely value for $E\{\bar{J}_{l,m}\}$ for a degree l in the range 13 to 19. Taking the root-square-sum of the coefficients in (6.15) and multiplying by 0.07×10^{-6}, we get $\pm 18.5 \times 10^{-6}$ for $E\{\Delta\lambda\}$, or about ± 140 meters. The observed value reported was about ± 100 meters, yielding a plausible 0.05×10^{-6} for the root-mean-square average of the $\bar{J}_{13,13}$ through $\bar{J}_{19,13}$. The contribution from $\bar{J}_{21,13}$, and so on should also be perceptible.

REFERENCES

1. Anderle, R. J. "Observations of Resonant Effects on Satellite Orbits Arising from the Thirteenth and Fourteenth-Order Tesseral Gravitational Coefficients." *J. Geophys. Res.*, *70* (1965), pp. 2453–2458.

2. Anderle, R. J. "Geodetic Parameters Set NWL-5E-6 Based on Doppler Satellite Observations." G. Veis, ed. *Proc. 2nd Int. Symp. Geod. on Use of Satellites*, Athens, in press, 1966.

3. Guier, W. H., and R. R. Newton. "The Earth's Gravity Field Deduced from the Doppler Tracking of Five Satellites." *J. Geophys. Res.*, *70* (1965), pp. 4613–4626.

4. Izsak, I. G. "A New Determination of Non-Zonal Harmonics by Satellites." J. Kovalevsky, ed. *Trajectories of Artificial Celestial Bodies as determined from Observations.* Berlin: Springer Verlag, in press, 1966.

5. King-Hele, D. G. and G. E. Cook. "The Even Zonal Harmonics in the Earth's Gravitational Potential." *Geophys. J.*, *10* (1965), pp. 17–30.

6. King-Hele, D. G., G. E. Cook, and D. W. Scott. "The Odd Zonal Harmonics in the Earth's Gravitational Potential." *Planet. Space Sci. 13* (1965), pp. 1213–1232.

7. Kozai, Y. "New Determination of Zonal Harmonic Coefficients in the Earth's Gravitational Potential." *Publ. Astron. Soc. Japan*, *16*, (1964), pp. 263–284.

8. Smith, D. E. "A Determination of the Odd Harmonics in the Geopotential Function." *Planet. Space Sci.*, *11* (1963), pp. 789–795.

9. Smith, D. E. "A Determination of the Even Harmonics in the Earth's Gravitational Potential Function." *Planet. Space Sci.*, *13* (1965), pp. 1151–1160.

10. Wagner, C. A. "A Determination of Earth's Equatorial Ellipticity from Seven Months of Syncom 2 Longitude Drift." *J. Geophys. Res.*, *70* (1965), pp. 1566–1568.

Index

INDEX

A CATALOG OF SELECTED

DOVER BOOKS

IN SCIENCE AND MATHEMATICS

THE HISTORICAL BACKGROUND OF CHEMISTRY, Henry M. Leicester. Evolution of ideas, not individual biography. Concentrates on formulation of a coherent set of chemical laws. 260pp. 5⅜ x 8½. 61053-5

A SHORT HISTORY OF CHEMISTRY, J. R. Partington. Classic exposition explores origins of chemistry, alchemy, early medical chemistry, nature of atmosphere, theory of valency, laws and structure of atomic theory, much more. 428pp. 5⅜ x 8½. (Available in U.S. only.) 65977-1

GENERAL CHEMISTRY, Linus Pauling. Revised 3rd edition of classic first-year text by Nobel laureate. Atomic and molecular structure, quantum mechanics, statistical mechanics, thermodynamics correlated with descriptive chemistry. Problems. 992pp. 5⅜ x 8½. 65622-5

Engineering

DE RE METALLICA, Georgius Agricola. The famous Hoover translation of greatest treatise on technological chemistry, engineering, geology, mining of early modern times (1556). All 289 original woodcuts. 638pp. 6¾ x 11. 60006-8

FUNDAMENTALS OF ASTRODYNAMICS, Roger Bate et al. Modern approach developed by U.S. Air Force Academy. Designed as a first course. Problems, exercises. Numerous illustrations. 455pp. 5⅜ x 8½. 60061-0

DYNAMICS OF FLUIDS IN POROUS MEDIA, Jacob Bear. For advanced students of ground water hydrology, soil mechanics and physics, drainage and irrigation engineering and more. 335 illustrations. Exercises, with answers. 784pp. 6⅛ x 9¼. 65675-6

ANALYTICAL MECHANICS OF GEARS, Earle Buckingham. Indispensable reference for modern gear manufacture covers conjugate gear-tooth action, gear-tooth profiles of various gears, many other topics. 263 figures. 102 tables. 546pp. 5⅜ x 8½. 65712-4

MECHANICS, J. P. Den Hartog. A classic introductory text or refresher. Hundreds of applications and design problems illuminate fundamentals of trusses, loaded beams and cables, etc. 334 answered problems. 462pp. 5⅜ x 8½. 60754-2

MECHANICAL VIBRATIONS, J. P. Den Hartog. Classic textbook offers lucid explanations and illustrative models, applying theories of vibrations to a variety of practical industrial engineering problems. Numerous figures. 233 problems, solutions. Appendix. Index. Preface. 436pp. 5⅜ x 8½. 64785-4

STRENGTH OF MATERIALS, J. P. Den Hartog. Full, clear treatment of basic material (tension, torsion, bending, etc.) plus advanced material on engineering methods, applications. 350 answered problems. 323pp. 5⅜ x 8½. 60755-0

A HISTORY OF MECHANICS, René Dugas. Monumental study of mechanical principles from antiquity to quantum mechanics. Contributions of ancient Greeks, Galileo, Leonardo, Kepler, Lagrange, many others. 671pp. 5⅜ x 8½. 65632-2

METAL FATIGUE, N. E. Frost, K. J. Marsh, and L. P. Pook. Definitive, clearly written, and well-illustrated volume addresses all aspects of the subject, from the historical development of understanding metal fatigue to vital concepts of the cyclic stress that causes a crack to grow. Includes 7 appendixes. 544pp. 5⅜ x 8½. 40927-9

STATISTICAL MECHANICS: Principles and Applications, Terrell L. Hill. Standard text covers fundamentals of statistical mechanics, applications to fluctuation theory, imperfect gases, distribution functions, more. 448pp. 5⅜ x 8½. 65390-0

THE VARIATIONAL PRINCIPLES OF MECHANICS, Cornelius Lanczos. Graduate level coverage of calculus of variations, equations of motion, relativistic mechanics, more. First inexpensive paperbound edition of classic treatise. Index. Bibliography. 418pp. 5⅜ x 8½. 65067-7

THE VARIOUS AND INGENIOUS MACHINES OF AGOSTINO RAMELLI: A Classic Sixteenth-Century Illustrated Treatise on Technology, Agostino Ramelli. One of the most widely known and copied works on machinery in the 16th century. 194 detailed plates of water pumps, grain mills, cranes, more. 608pp. 9 x 12. 28180-9

ORDINARY DIFFERENTIAL EQUATIONS AND STABILITY THEORY: An Introduction, David A. Sánchez. Brief, modern treatment. Linear equation, stability theory for autonomous and nonautonomous systems, etc. 164pp. 5⅜ x 8¼. 63828-6

ROTARY WING AERODYNAMICS, W. Z. Stepniewski. Clear, concise text covers aerodynamic phenomena of the rotor and offers guidelines for helicopter performance evaluation. Originally prepared for NASA. 537 figures. 640pp. 6⅛ x 9¼. 64647-5

INTRODUCTION TO SPACE DYNAMICS, William Tyrrell Thomson. Comprehensive, classic introduction to space-flight engineering for advanced undergraduate and graduate students. Includes vector algebra, kinematics, transformation of coordinates. Bibliography. Index. 352pp. 5⅜ x 8½. 65113-4

HISTORY OF STRENGTH OF MATERIALS, Stephen P. Timoshenko. Excellent historical survey of the strength of materials with many references to the theories of elasticity and structure. 245 figures. 452pp. 5⅜ x 8½. 61187-6

ANALYTICAL FRACTURE MECHANICS, David J. Unger. Self-contained text supplements standard fracture mechanics texts by focusing on analytical methods for determining crack-tip stress and strain fields. 336pp. 6⅛ x 9¼. 41737-9

Mathematics

HANDBOOK OF MATHEMATICAL FUNCTIONS WITH FORMULAS, GRAPHS, AND MATHEMATICAL TABLES, edited by Milton Abramowitz and Irene A. Stegun. Vast compendium: 29 sets of tables, some to as high as 20 places. 1,046pp. 8 x 10½. 61272-4

THE FUNCTIONS OF MATHEMATICAL PHYSICS, Harry Hochstadt. Comprehensive treatment of orthogonal polynomials, hypergeometric functions, Hill's equation, much more. Bibliography. Index. 322pp. 5⅜ x 8½. 65214-9

THREE PEARLS OF NUMBER THEORY, A. Y. Khinchin. Three compelling puzzles require proof of a basic law governing the world of numbers. Challenges concern van der Waerden's theorem, the Landau-Schnirelmann hypothesis and Mann's theorem, and a solution to Waring's problem. Solutions included. 64pp. 5⅜ x 8½. 40026-3

CALCULUS REFRESHER FOR TECHNICAL PEOPLE, A. Albert Klaf. Covers important aspects of integral and differential calculus via 756 questions. 566 problems, most answered. 431pp. 5⅜ x 8½. 20370-0

THE PHILOSOPHY OF MATHEMATICS: An Introductory Essay, Stephan Körner. Surveys the views of Plato, Aristotle, Leibniz & Kant concerning propositions and theories of applied and pure mathematics. Introduction. Two appendices. Index. 198pp. 5⅜ x 8½. 25048-2

INTRODUCTORY REAL ANALYSIS, A.N. Kolmogorov, S. V. Fomin. Translated by Richard A. Silverman. Self-contained, evenly paced introduction to real and functional analysis. Some 350 problems. 403pp. 5⅜ x 8½. 61226-0

APPLIED ANALYSIS, Cornelius Lanczos. Classic work on analysis and design of finite processes for approximating solution of analytical problems. Algebraic equations, matrices, harmonic analysis, quadrature methods, much more. 559pp. 5⅜ x 8½. 65656-X

AN INTRODUCTION TO ALGEBRAIC STRUCTURES, Joseph Landin. Superb self-contained text covers "abstract algebra": sets and numbers, theory of groups, theory of rings, much more. Numerous well-chosen examples, exercises. 247pp. 5⅜ x 8½. 65940-2

SPECIAL FUNCTIONS, N. N. Lebedev. Translated by Richard Silverman. Famous Russian work treating more important special functions, with applications to specific problems of physics and engineering. 38 figures. 308pp. 5⅜ x 8½. 60624-4

QUALITATIVE THEORY OF DIFFERENTIAL EQUATIONS, V. V. Nemytskii and V.V. Stepanov. Classic graduate-level text by two prominent Soviet mathematicians covers classical differential equations as well as topological dynamics and ergodic theory. Bibliographies. 523pp. 5⅜ x 8½. 65954-2

NUMBER THEORY AND ITS HISTORY, Oystein Ore. Unusually clear, accessible introduction covers counting, properties of numbers, prime numbers, much more. Bibliography. 380pp. 5⅜ x 8½. 65620-9

THEORY OF MATRICES, Sam Perlis. Outstanding text covering rank, nonsingularity and inverses in connection with the development of canonical matrices under the relation of equivalence, and without the intervention of determinants. Includes exercises. 237pp. 5⅜ x 8½. 66810-X

INTRODUCTION TO ANALYSIS, Maxwell Rosenlicht. Unusually clear, accessible coverage of set theory, real number system, metric spaces, continuous functions, Riemann integration, multiple integrals, more. Wide range of problems. Undergraduate level. Bibliography. 254pp. 5⅜ x 8½. 65038-3

MODERN NONLINEAR EQUATIONS, Thomas L. Saaty. Emphasizes practical solution of problems; covers seven types of equations. ". . . a welcome contribution to the existing literature...."–*Math Reviews*. 490pp. 5⅜ x 8½. 64232-1

MATRICES AND LINEAR ALGEBRA, Hans Schneider and George Phillip Barker. Basic textbook covers theory of matrices and its applications to systems of linear equations and related topics such as determinants, eigenvalues and differential equations. Numerous exercises. 432pp. 5⅜ x 8½. 66014-1

MATHEMATICS APPLIED TO CONTINUUM MECHANICS, Lee A. Segel. Analyzes models of fluid flow and solid deformation. For upper-level math, science and engineering students. 608pp. 5⅜ x 8½. 65369-2

ELEMENTS OF REAL ANALYSIS, David A. Sprecher. Classic text covers fundamental concepts, real number system, point sets, functions of a real variable, Fourier series, much more. Over 500 exercises. 352pp. 5⅜ x 8½. 65385-4

AN INTRODUCTION TO MATRICES, SETS AND GROUPS FOR SCIENCE STUDENTS, G. Stephenson. Concise, readable text introduces sets, groups, and most importantly, matrices to undergraduate students of physics, chemistry, and engineering. Problems. 164pp. 5⅜ x 8½. 65077-4

SET THEORY AND LOGIC, Robert R. Stoll. Lucid introduction to unified theory of mathematical concepts. Set theory and logic seen as tools for conceptual understanding of real number system. 496pp. 5⅜ x 8¼. 63829-4

TENSOR CALCULUS, J.L. Synge and A. Schild. Widely used introductory text covers spaces and tensors, basic operations in Riemannian space, non-Riemannian spaces, etc. 324pp. 5⅜ x 8¼. 63612-7

ORDINARY DIFFERENTIAL EQUATIONS, Morris Tenenbaum and Harry Pollard. Exhaustive survey of ordinary differential equations for undergraduates in mathematics, engineering, science. Thorough analysis of theorems. Diagrams. Bibliography. Index. 818pp. 5⅜ x 8½. 64940-7

INTEGRAL EQUATIONS, F. G. Tricomi. Authoritative, well-written treatment of extremely useful mathematical tool with wide applications. Volterra Equations, Fredholm Equations, much more. Advanced undergraduate to graduate level. Exercises. Bibliography. 238pp. 5⅜ x 8½. 64828-1

FOURIER SERIES, Georgi P. Tolstov. Translated by Richard A. Silverman. A valuable addition to the literature on the subject, moving clearly from subject to subject and theorem to theorem. 107 problems, answers. 336pp. 5⅜ x 8½. 63317-9

POPULAR LECTURES ON MATHEMATICAL LOGIC, Hao Wang. Noted logician's lucid treatment of historical developments, set theory, model theory, recursion theory and constructivism, proof theory, more. 3 appendixes. Bibliography. 1981 edition. ix + 283pp. 5⅜ x 8½. 67632-3

CALCULUS OF VARIATIONS, Robert Weinstock. Basic introduction covering isoperimetric problems, theory of elasticity, quantum mechanics, electrostatics, etc. Exercises throughout. 326pp. 5⅜ x 8½. 63069-2

THE CONTINUUM: A Critical Examination of the Foundation of Analysis, Hermann Weyl. Classic of 20th-century foundational research deals with the conceptual problem posed by the continuum. 156pp. 5⅜ x 8½. 67982-9

CHALLENGING MATHEMATICAL PROBLEMS WITH ELEMENTARY SOLUTIONS, A. M. Yaglom and I. M. Yaglom. Over 170 challenging problems on probability theory, combinatorial analysis, points and lines, topology, convex polygons, many other topics. Solutions. Total of 445pp. 5⅜ x 8½. Two-vol. set.
Vol. I: 65536-9 Vol. II: 65537-7

A SURVEY OF NUMERICAL MATHEMATICS, David M. Young and Robert Todd Gregory. Broad self-contained coverage of computer-oriented numerical algorithms for solving various types of mathematical problems in linear algebra, ordinary and partial, differential equations, much more. Exercises. Total of 1,248pp. 5⅜ x 8½. Two volumes. Vol. I: 65691-8 Vol. II: 65692-6

INTRODUCTION TO PARTIAL DIFFERENTIAL EQUATIONS WITH APPLICATIONS, E. C. Zachmanoglou and Dale W. Thoe. Essentials of partial differential equations applied to common problems in engineering and the physical sciences. Problems and answers. 416pp. 5⅜ x 8½. 65251-3

THE THEORY OF GROUPS, Hans J. Zassenhaus. Well-written graduate-level text acquaints reader with group-theoretic methods and demonstrates their usefulness in mathematics. Axioms, the calculus of complexes, homomorphic mapping, *p*-group theory, more. Many proofs shorter and more transparent than older ones. 276pp. 5⅜ x 8½. 40922-8

DISTRIBUTION THEORY AND TRANSFORM ANALYSIS: An Introduction to Generalized Functions, with Applications, A. H. Zemanian. Provides basics of distribution theory, describes generalized Fourier and Laplace transformations. Numerous problems. 384pp. 5⅜ x 8½. 65479-6

Math–Decision Theory, Statistics, Probability

ELEMENTARY DECISION THEORY, Herman Chernoff and Lincoln E. Moses. Clear introduction to statistics and statistical theory covers data processing, probability and random variables, testing hypotheses, much more. Exercises. 364pp. 5⅜ x 8½. 65218-1

Math–Geometry and Topology

ELEMENTARY CONCEPTS OF TOPOLOGY, Paul Alexandroff. Elegant, intuitive approach to topology from set-theoretic topology to Betti groups; how concepts of topology are useful in math and physics. 25 figures. 57pp. 5⅜ x 8½. 60747-X

COMBINATORIAL TOPOLOGY, P. S. Alexandrov. Clearly written, well-organized, three-part text begins by dealing with certain classic problems without using the formal techniques of homology theory and advances to the central concept, the Betti groups. Numerous detailed examples. 654pp. 5⅜ x 8½. 40179-0

EXPERIMENTS IN TOPOLOGY, Stephen Barr. Classic, lively explanation of one of the byways of mathematics. Klein bottles, Moebius strips, projective planes, map coloring, problem of the Koenigsberg bridges, much more, described with clarity and wit. 43 figures. 210pp. 5⅜ x 8½. 25933-1

CONFORMAL MAPPING ON RIEMANN SURFACES, Harvey Cohn. Lucid, insightful book presents ideal coverage of subject. 334 exercises make book perfect for self-study. 55 figures. 352pp. 5⅝ x 8¼. 64025-6

THE GEOMETRY OF RENÉ DESCARTES, René Descartes. The great work founded analytical geometry. Original French text, Descartes's own diagrams, together with definitive Smith-Latham translation. 244pp. 5⅜ x 8½. 60068-8

THE THIRTEEN BOOKS OF EUCLID'S ELEMENTS, translated with introduction and commentary by Sir Thomas L. Heath. Definitive edition. Textual and linguistic notes, mathematical analysis. 2,500 years of critical commentary. Unabridged. 1,414pp. 5⅜ x 8½. Three-vol. set.

Vol. I: 60088-2 Vol. II: 60089-0 Vol. III: 60090-4

GEOMETRY OF COMPLEX NUMBERS, Hans Schwerdtfeger. Illuminating, widely praised book on analytic geometry of circles, the Moebius transformation, and two-dimensional non-Euclidean geometries. 200pp. 5⅝ x 8¼. 63830-8

DIFFERENTIAL GEOMETRY, Heinrich W. Guggenheimer. Local differential geometry as an application of advanced calculus and linear algebra. Curvature, transformation groups, surfaces, more. Exercises. 62 figures. 378pp. 5⅜ x 8½. 63433-7

CURVATURE AND HOMOLOGY: Enlarged Edition, Samuel I. Goldberg. Revised edition examines topology of differentiable manifolds; curvature, homology of Riemannian manifolds; compact Lie groups; complex manifolds; curvature, homology of Kaehler manifolds. New Preface. Four new appendixes. 416pp. 5⅜ x 8½. 40207-X

TOPOLOGY, John G. Hocking and Gail S. Young. Superb one-year course in classical topology. Topological spaces and functions, point-set topology, much more. Examples and problems. Bibliography. Index. 384pp. 5⅝ x 8¼. 65676-4

LECTURES ON CLASSICAL DIFFERENTIAL GEOMETRY, Second Edition, Dirk J. Struik. Excellent brief introduction covers curves, theory of surfaces, fundamental equations, geometry on a surface, conformal mapping, other topics. Problems. 240pp. 5⅜ x 8½. 65609-8

Math–History of

A SHORT ACCOUNT OF THE HISTORY OF MATHEMATICS, W. W. Rouse Ball. One of clearest, most authoritative surveys from the Egyptians and Phoenicians through 19th-century figures such as Grassman, Galois, Riemann. Fourth edition. 522pp. 5⅜ x 8½. 20630-0

THE HISTORY OF THE CALCULUS AND ITS CONCEPTUAL DEVELOPMENT, Carl B. Boyer. Origins in antiquity, medieval contributions, work of Newton, Leibniz, rigorous formulation. Treatment is verbal. 346pp. 5⅜ x 8½. 60509-4

THE HISTORICAL ROOTS OF ELEMENTARY MATHEMATICS, Lucas N. H. Bunt, Phillip S. Jones, and Jack D. Bedient. Fundamental underpinnings of modern arithmetic, algebra, geometry and number systems derived from ancient civilizations. 320pp. 5⅜ x 8½. 25563-8

A HISTORY OF MATHEMATICAL NOTATIONS, Florian Cajori. This classic study notes the first appearance of a mathematical symbol and its origin, the competition it encountered, its spread among writers in different countries, its rise to popularity, its eventual decline or ultimate survival. Original 1929 two-volume edition presented here in one volume. xxviii+820pp. 5⅜ x 8½. 67766-4

GAMES, GODS & GAMBLING: A History of Probability and Statistical Ideas, F. N. David. Episodes from the lives of Galileo, Fermat, Pascal, and others illustrate this fascinating account of the roots of mathematics. Features thought-provoking references to classics, archaeology, biography, poetry. 1962 edition. 304pp. 5⅜ x 8½. (Available in U.S. only.) 40023-9

OF MEN AND NUMBERS: The Story of the Great Mathematicians, Jane Muir. Fascinating accounts of the lives and accomplishments of history's greatest mathematical minds–Pythagoras, Descartes, Euler, Pascal, Cantor, many more. Anecdotal, illuminating. 30 diagrams. Bibliography. 256pp. 5⅜ x 8½. 28973-7

HISTORY OF MATHEMATICS, David E. Smith. Nontechnical survey from ancient Greece and Orient to late 19th century; evolution of arithmetic, geometry, trigonometry, calculating devices, algebra, the calculus. 362 illustrations. 1,355pp. 5⅜ x 8½. Two-vol. set. Vol. I: 20429-4 Vol. II: 20430-8

A CONCISE HISTORY OF MATHEMATICS, Dirk J. Struik. The best brief history of mathematics. Stresses origins and covers every major figure from ancient Near East to 19th century. 41 illustrations. 195pp. 5⅜ x 8½. 60255-9

Physics

OPTICAL RESONANCE AND TWO-LEVEL ATOMS, L. Allen and J. H. Eberly. Clear, comprehensive introduction to basic principles behind all quantum optical resonance phenomena. 53 illustrations. Preface. Index. 256pp. 5⅜ x 8½. 65533-4

ULTRASONIC ABSORPTION: An Introduction to the Theory of Sound Absorption and Dispersion in Gases, Liquids and Solids, A. B. Bhatia. Standard reference in the field provides a clear, systematically organized introductory review of fundamental concepts for advanced graduate students, research workers. Numerous diagrams. Bibliography. 440pp. 5⅜ x 8½. 64917-2

QUANTUM THEORY, David Bohm. This advanced undergraduate-level text presents the quantum theory in terms of qualitative and imaginative concepts, followed by specific applications worked out in mathematical detail. Preface. Index. 655pp. 5⅜ x 8½. 65969-0

ATOMIC PHYSICS (8th edition), Max Born. Nobel laureate's lucid treatment of kinetic theory of gases, elementary particles, nuclear atom, wave-corpuscles, atomic structure and spectral lines, much more. Over 40 appendices, bibliography. 495pp. 5⅜ x 8½. 65984-4

AN INTRODUCTION TO HAMILTONIAN OPTICS, H. A. Buchdahl. Detailed account of the Hamiltonian treatment of aberration theory in geometrical optics. Many classes of optical systems defined in terms of the symmetries they possess. Problems with detailed solutions. 1970 edition. xv + 360pp. 5⅜ x 8½. 67597-1

THIRTY YEARS THAT SHOOK PHYSICS: The Story of Quantum Theory, George Gamow. Lucid, accessible introduction to influential theory of energy and matter. Careful explanations of Dirac's anti-particles, Bohr's model of the atom, much more. 12 plates. Numerous drawings. 240pp. 5⅜ x 8½. 24895-X

ELECTRONIC STRUCTURE AND THE PROPERTIES OF SOLIDS: The Physics of the Chemical Bond, Walter A. Harrison. Innovative text offers basic understanding of the electronic structure of covalent and ionic solids, simple metals, transition metals and their compounds. Problems. 1980 edition. 582pp. 6⅛ x 9¼. 66021-4

HYDRODYNAMIC AND HYDROMAGNETIC STABILITY, S. Chandrasekhar. Lucid examination of the Rayleigh-Benard problem; clear coverage of the theory of instabilities causing convection. 704pp. 5⅜ x 8¼. 64071-X

INVESTIGATIONS ON THE THEORY OF THE BROWNIAN MOVEMENT, Albert Einstein. Five papers (1905–8) investigating dynamics of Brownian motion and evolving elementary theory. Notes by R. Fürth. 122pp. 5⅜ x 8½. 60304-0

THE PHYSICS OF WAVES, William C. Elmore and Mark A. Heald. Unique overview of classical wave theory. Acoustics, optics, electromagnetic radiation, more. Ideal as classroom text or for self-study. Problems. 477pp. 5⅜ x 8½. 64926-1

Princeton U. Store Clearance
New Jersey
Thu 25 Oct 2007
$3.98 + .28 tax
$9.95 list